The Trouble with Ancient DNA

The Trouble with Ancient DNA

Telling Stories of the Past with Genomic Science

ANNA KÄLLÉN

The University of Chicago Press
Chicago and London

The University of Chicago Press, Chicago 60637
The University of Chicago Press, Ltd., London
© 2025 by The University of Chicago
Published 2025
Printed in the United States of America

34 33 32 31 30 29 28 27 26 25   1 2 3 4 5

ISBN-13: 978-0-226-82167-2 (cloth)
ISBN-13: 978-0-226-83557-0 (paper)
ISBN-13: 978-0-226-83556-3 (e-book)
DOI: https://doi.org/10.7208/chicago/9780226835563.001.0001

Library of Congress Cataloging-in-Publication Data

Names: Källén, Anna, author.
Title: The trouble with ancient DNA : telling stories of the past with genomic science / Anna Källén.
Description: Chicago : The University of Chicago Press, 2025. | Includes bibliographical references and index.
Identifiers: LCCN 2024021584 | ISBN 9780226821672 (cloth) | ISBN 9780226835570 (paperback) | ISBN 9780226835563 (ebook)
Subjects: LCSH: DNA, Fossil. | Genomics.
Classification: LCC QP624 .K37 2025 | DDC 572/.786—dc23/eng/20240607
LC record available at https://lccn.loc.gov/2024021584

♾ This paper meets the requirements of ANSI/NISO Z39.48-1992 (Permanence of Paper).

Contents

Introduction

Working with prehistoric epochs gives the scientist many opportunities to realize their own slightness and inability, and this in more than one way.
—Hanna Rydh, 1934[1]

It was a brisk but ordinary Thursday in September 2017. The fall semester had just begun, and the morning bus was packed with students on the way to their introductory lectures. Across the road from my office at Stockholm University, my colleagues in the Department of Archaeology arrived to another day at work, unaware of the storm that was coming their way.

"DNA proves fearsome Viking warrior was a woman."[2] The story hit the headlines almost immediately and spread with the speed of light. News features with exclamation marks were soon seen in popular media all over the world. *Science* magazine reported on new genetic evidence of a formidable woman, and the *Daily Mail* used all-capital letters for the key words: "First Proof There Really Were FEMALE Viking Warriors: DNA Study Reveals High-ranking 'Valkyrie' Buried with Weapons and Two Horses Was a Five Foot Six WOMAN."[3]

The cause of the excitement was an article published that very day in the "Brief Communications" section of the *American Journal of Physical Anthropology*. It had the title "A Female Viking Warrior Confirmed by Genomics"[4] and was authored by my colleagues across the road. They were experts on Viking Age society and had known for a long time that the skeleton that the article was about—which was uncovered in 1878 from an extraordinary grave with horses and typical warrior equipment—was female. It was quite clear from the size of the bones and shape of the pelvis, and two academic papers had already claimed this as a fact. Hence, for them, the result of the DNA analysis that determined the absence of a Y chromosome came as no surprise.

My colleagues were by no means shy about media attention. On the contrary, they were well aware of the popular value of Viking stories, and some of them had collaborated with the production team of the TV series *Vikings* as a way to broaden the reach of their research.[5] Despite this, they had underestimated the explosive power of a cocktail that mixed Vikings with DNA, and stood unprepared for the tsunami of media attention that flowed over them. As the days, weeks, and months passed, they tried to stay afloat while contributing to interviews and adding details about the female Viking warrior's characteristics and professional capacities. Almost all of the stories were accompanied by illustrations of young, white models with beautifully braided blond hair and serious expressions, sporting tight garments and striking poses in more or less violent situations. Of course, no such traits were evident in the analysis of DNA molecules, which merely confirmed the absence of a Y chromosome in the remains found in the grave.[6] Nevertheless, the bold headlines spoke of DNA as irrefutable evidence—not only as proof that the individual buried in that particular grave was chromosomally female, but more generally as proof of the high social status and military rank of the buried person, and of the common existence of fierce and fearsome warrior women leading Viking armies on the battlefield. Before the furor eventually calmed, the female Viking warrior had been featured in innumerable news articles, several international film productions, various artistic reconstructions, a dedicated porn site with reference to the original research paper, and a plethora of Twitter (X) threads that even reached the corridors of the US White House.[7]

Meanwhile, in the United States, other developments in ancient DNA research were about to make headlines. These had been simmering for some time. Already in 2015, an article in *Science* magazine covered the latest developments in ancient DNA research under the title "**REVOLUTION IN HUMAN EVOLUTION**," set in bold capital letters. "The whole field is exploding in terms of its impact," said one geneticist interviewed. "The data that's coming out is completely rewriting what we know about human prehistory." A few months after the headlines about the female Viking warrior, another story, speaking of a revolution in ancient DNA research, came along to stir the debate. This time it was David Reich, a professor of population genetics and vocal leader of a laboratory at Harvard University, who inspired headlines with his new book, *Who We Are and How We Got Here*. More specifically, the debate came to circle around an op-ed in the *New York Times* where Reich presented his view on the concept of race. "As a geneticist," he wrote, "[I] know that it is simply no longer possible to ignore average genetic differences among 'races.'"[8] The statement cracked open a rift through the scientific community. A forceful counterargument, written and signed by sixty-seven prominent scholars and scientists, debunked the statement as "seriously flawed,"[9] but there were also plenty of commentators applauding and endorsing Reich's arguments.

From the other side of the Atlantic, I followed these debates with intense interest. Having observed the story of the female Viking warrior, with the abyss of distance between the information yielded by the DNA analysis and the sensational images and articles it inspired, I began to sense that ancient DNA research was not just about molecules, but also about signs and stories. I was particularly struck by a sentence in the *New York Times* op-ed, where a seemingly concerned Reich wrote that he was "worried that well-meaning people [. . .] are digging themselves into an indefensible position, one that will not survive the onslaught of science."[10] What was the deeper significance of this sentence? What precisely was the "onslaught of science" awaiting those who acknowledged the great complexities of human life and human relations? There began my own inquiries into the rhetoric and reality of ancient DNA research.

⋊⋉

When I was twenty-one years old, I traveled to Southeast Asia with a friend. We were students of archaeology on a Swedish foreign-aid mission to help the people of Laos to gain knowledge of their ancient past. Or so we thought. We surveyed, excavated, and wrote archaeological reports. Years later, I had gained some knowledge of the ancient layers of the sites where we worked. Even more, I had begun to see how ignorant I was. Rather than being helped, the people in Laos had helped me, by challenging my worldview, over and over again. Spending years as a foreign barbarian with language skills that could not distinguish "I" from "penis," among intelligent people who already knew their past (albeit with other ways of thinking about time, material things, and spiritual presence), made me realize the slightness and partiality of my own archaeological perspective. Quite a few slices of humble pie went down in those years.

Ever since, I have been interested in the structures and premises of archaeological storytelling. How did archaeology come to dominate official knowledge about the ancient past, and what have been the effects of that dominance? I have never doubted or questioned the facts underlying archaeological stories of the past, but I have become aware of just how fragile and flexible those facts are, and how easily they can be bent to fit stories that serve the present. Over the course of the late nineteenth and early twentieth centuries, prehistoric people around the world have been squeezed into nuclear families and ethnic communities featured in teleological stories of development through Stone, Bronze, and Iron Ages, to explain and motivate social hierarchies and political decisions in colonial projects and modern nation-states. With hindsight and analytical distance, the contemporary contingencies and consequences of such stories stand out clearly, but from the perspective of the storytellers, they likely appeared neutral and true, just as the history we write appears neutral and true to us.

Since the beginning of my own critical reflections, on my work in Laos and on archaeological research and storytelling more broadly, I have taken much inspiration and important analytical concepts from science and culture theorists like Donna Haraway, James Clifford, and Bruno Latour.[11] Their writings have shown how works in science and anthropology are always connected with their societal contexts, and are thus contingent upon the languages, images, technologies, historical knowledges, social hierarchies, economic structures, and political power relations of their practitioners and their moment in time. For me, this has become a fundamental perspective, one that you will find at the heart of this book. I have combined it, here and elsewhere, with a burning interest in the politics of the past. Here I have found grounding in the works of postcolonial theorists like Trinh T. Minh-ha, Edward Said, and Homi Bhabha.[12] They approach national and ethnic identity not as a timeless essential quality that can be uncovered in bodies or things, but rather as a form of cultural production that involves bodies, things, and ideas—a process through which groups are shaped and re-shaped in historically contingent yet creative processes of negotiation. For my own studies, this has meant that I see ethnic identity or other group qualities in the ancient past as results of creative processes of negotiation, in the past as well as in the present. I have also assumed this perspective to explore how the identities and meaningful narratives we ascribe to people and things in the past tend to come back and affect us in the present in politically useful, and sometimes harmful, ways.

I have by no means been alone with these perspectives. Over the thirty years that have passed since I first set foot in Laos, the general awareness of academic uses and abuses of the past has grown exponentially. Uses of the past have been studied by critical anthropologists, archaeologists, and historians, in critical heritage studies, science and technology studies, and the philosophy of science. The historical lessons learned from the uses of the past to justify nineteenth-century colonial oppression and twentieth-century genocide in World War II have been complemented by more recent postcolonial, feminist, and Indigenous perspectives. Together, they paved the way for a humbler archaeology that acknowledged its own fragility as an ultimate guessing game and opened up broader perspectives, in the past as well as in the present.

ↄ◁▯▷ↄ

Then came the "ancient DNA revolution." From 2015 onward, it inspired a new level of storytelling with steadfast claims to have uncovered the exact identities of people in the ancient past. Both geneticists and archaeologists seized the opportunity to splash out on metaphors, and they painted vivid pictures of ancient societies with muscular murderous men and fierce women. There were stories of groups of people blasting across continents to become our own ancestors—all apparently confirmed by new genomic science. In these stories, presented with evangelical enthusiasm, genomic science played the role of an all-seeing God's eye—a wonderful new machine with the ability to reveal the true identities of people in the ancient past. The popular science media gulped up the messages and pumped up the volume, and soon we were deluged with strong images and resolute stories claiming to have cracked archaeological mysteries and settled long-standing controversies, once and for all. The few calls for caution that were heard were dismissed as anxious, jealous, or ignorant of the possibilities of the new genomic science.

When the dramatic stories first appeared, I was astounded by their boldness and their claims to absolute knowledge of ancient people's identities. From a distance they appeared clearly tainted by the identitarianism that is characteristic of our time, and I could sense a risk of dangerous consequences as they planted absolute notions of ethnic and national identity in prehistoric times. They seemed in many ways outmoded in a time when the political climate called for caution and sensibility, and when the need for trustworthy science seemed more acute than ever. But then I had to admit that I was indeed ignorant of the possibilities of the new genomic science.

To learn more, I teamed up with three other researchers: geneticist Charlotte Mulcare, media historian Andreas Nyblom, and historian of ideas Daniel Strand. We have worked together since 2018 to gain more profound knowledge and understanding of the global developments in archaeogenomics. I have learned so much from them all, but having Charlotte on the team broadened my horizons in ways I could not foresee. With outstanding pedagogical skills and endless patience, she has translated the technical jargon, statistics, tables, and diagrams of scientific papers and supplementary materials so I can properly appreciate what genomic analyses can and cannot do.

If my initial reaction was typical of someone with a social-constructivist perspective—that the grand stories of ancient DNA could not live up to their claims of total vision of the past, because there is no such thing as total vision, and we know for a fact that people conceive of identity, kinship, and materiality in ways that are not restricted to the reductive and conclusive realms of genetic science—working with Charlotte has offered insight into the black box of genetics.[13] I now know that genetic analyses are not neutral but creative, formatting the ancient past in specific figures that have just as much to do with the structures of computer programs and the imaginations of those who do the programming as with the ancient past. For me, this was a revelation, especially in light of the popular and political potency of widespread stories based on ancient DNA. It has convinced me of the need to think carefully about the potentials and pitfalls of telling stories of the ancient past with genomic science, *before* we pick up the megaphone.

From the work my colleagues and I have done together, I have also learned that archaeologists and genetic scientists are two diverse groups of individuals with a variety of ideas and perspectives. The "ancient DNA revolution" has been accompanied by considerable boasting that DNA is the solution to any big question in archaeology—and not only from genetic scientists. Archaeologists, leading science journals, and the popular media have contributed just as much to the hype. At the same time, a number of archaeologists and historians, as well as genetic scientists and molecular anthropologists, have called for caution and clearly explained the pitfalls of using DNA technology to research and establish historic identity. In the United States and other parts of the world, there have been important initiatives by genetic scientists to highlight ethical issues in dealing with human remains and to collaborate with Indigenous communities in studies of ancient DNA. I will offer more details and examples further on, but I wish to clarify here that it is the highly visible and influential *storytelling* associated with ancient DNA, and not the scientists or projects working with it, that is my primary subject of critique in this book.

That said, one key argument runs through all of the chapters: if we ignore the potential predicaments and perils of writing history, the generic structures and models of genetic science will tend to shape the ancient past into certain forms, figures, and narratives that can have dangerous consequences in politically sensitive contexts. Moreover, when it comes to telling stories about the past with authority, the great *symbolic* value of DNA as evidence has elevated genetic science to a privileged position with respect to more nuanced and critically informed cultural interpretations—not least among the general public. Most researchers working with ancient DNA are primarily interested in methodological development, and operate in restricted local contexts with fine-grained analyses and knowledge claims that are balanced and appropriate to the questions and material of their studies. However, a few scientists and scholars who have exploited the symbolic value of DNA to tell sensational stories have been rewarded with major research grants, prizes, and public visibility. And those evangelical claims that DNA offers a perfect window onto the ancient past have gained purchase in the popular press and have had the greatest impact on broader audiences.[14]

This has led me to the insight that we all have more to learn. The vast majority of archaeologists, journalists, and the interested public who consume sensational media stories about ancient DNA have a poor understanding of genetic methodology. Likewise, and equally important, the knowledge of historical research and storytelling is poor among most geneticists. This creates a situation where the parties involved in ancient DNA research tend to caricature one another. In this situation, archaeologists and the interested public may treat DNA as simple evidence, as seen in the latest episode of *CSI*,[15] and geneticists may see history-writing as a lightweight pursuit requiring few skills other than a general interest and writing talent. At best, such a situation of mutual misapprehension will cause problems for the scientists and scholars involved. At worst, it will contribute to the telling of dangerous stories with serious consequences. Hence, there is every reason for all parties involved to learn more about one another, with the ultimate goal of finding ways to work with ancient DNA that allow us to learn interesting new things about the ancient past, with as little harm as possible. This book aims to make a contribution to that end. But do not expect to find here an instruction manual or a complete set of practical guidelines. See it rather as a travel companion—a critical friend to hold your hand and alert you to potential pitfalls—as you explore the slippery paths of writing history with genomic science.

✕⬭✕

We begin with a background to ancient DNA as an object of research and popular imagination, with particular focus on the so-called revolution in ancient DNA. The aim this first chapter is to provide basic information on what the DNA molecule can and cannot do when we tell stories of the ancient past. The following three chapters delve deep into some of the most widespread stories of ancient DNA, teasing out their contingencies and discussing their real and potential consequences. Chapter 2, "Return of the Arrows," discusses some remarkable stories of prehistoric migration based on new studies of ancient DNA and the notions of national and ethnic identity they have awakened and inspired. Chapter 3, "A Family Tree of Everyone," explores how images and ideas of genetic ancestry connect people of the past and present with social and political implications. Chapter 4, "Paleopersonalities," homes in on the role of genetic profiling in the making of prehistoric personalities, with particular focus on the story of Cheddar Man, "Mesolithic Britain's blue-eyed boy." The bottom line, which will shine through most clearly in the concluding chapter, is that there is good reason to rein back to a more sensible and knowledgeable position. It is time to unlearn the evangelical rhetoric of the "ancient DNA revolution" and its truth-machine image of genomic science. Only then can we learn more about the real potentials, and perils, of using ancient DNA to tell stories about the past.

1: Ancient DNA

Until the 1990s, archaeological sites provided much of the evidence for the spread of people across the Americas. [. . .] Yet archaeology alone has left many questions unanswered, such as who *exactly* lived in those early sites and how they were related to each other. Geneticists are seeking to answer some of those questions by looking at the DNA of living Native Americans.[1]

This sounds reasonable, right? You have probably heard it before, the story of how archaeology has hitherto only provided a fragmentary vision of the past, and how DNA, as a secure source of clear answers, has offered closure on long-standing issues and debates, such as the question of who were the first inhabitants of the Americas.

But take a moment and ask yourself—who are you, exactly? If I were to answer, I would say that I am a mother of three beautiful children, two stepchildren and one by birth. I am a trained archaeologist working as a professor of museology at Umeå University, 500 kilometers (about 310 miles) north of Stockholm, where I live with my partner of twenty years. Although I was born and now live and work on the east coast of Sweden, I have a distinct native accent from my childhood on the west coast. I love to host a cocktail party just as much as I love to be alone, foraging for mushrooms in the forest. I am tall, quite a lot taller than the national average and taller than the rest of my birth family. I am a Swedish citizen but have worked in different parts of the world, and much of my adult life and identity has been defined by my work in Laos in Southeast Asia, which for many years I thought of as my second home.

All right, I had better stop there. You might have found other aspects more important for a description of your own identity, such as religion, race, ethnicity, caste, class, tribe, gender, favorite football team, medical conditions, where you went to university, or your political party affiliation. Details aside, I suppose your answers sound pretty much like mine. They are complicated, coded by language and culture, and almost every sentence involves multiple, sometimes contradictory, aspects of our personal identities. They would not have been the same had the question been put to us twenty years ago, and they will not be the same twenty years from now. In short: human identities are always complicated and in flux, at times contradictory, and can only be described from situated positions.

Now, what of the above—my own answer to who I am, exactly— would be possible to trace in an analysis of my DNA? The truth is, almost nothing. A DNA analysis could tell that I am chromosomally female with a birth family connection to people now living in Scandinavia. If the analysis was made from my remains, say, a thousand years from now and included some of my close descendants, it could possibly say that I had offspring with a male also connected to Scandinavia. So, a story about me based on analyses of my own and my descendants' genomes would conclude that I lived my life much in the exact same place where I was born and that I had a child with a local male. On an individual level, it could possibly have indicated that I was likely to have been above the average height—but not exactly how tall. It would probably include a guess at the color of my skin, hair, and eyes, and whether I have a predisposition for waxy ears or hereditary diseases of which I am blissfully unaware.[2]

A DNA analysis, as far as it goes today, can indeed give some information about our physical beings, if we know where to look and how to ask the right questions. Importantly, however, the results of a DNA analysis would contain close to nothing of all that I have just told you about who I am. Genetic science alone can say nothing about close family relations that are not based on biological reproduction, such as my stepchildren. It cannot reveal travels and movements, such as my time in Laos, unless they have resulted in biological offspring. It can say nothing about interests and sentiments, languages and accents, or professional occupations—what constitutes our identities and alliances today, and likely constituted the identities and alliances of people in the past. Osteological analyses of my bones and archaeological investigations of my home and my grave would probably get much closer. And yet we hear, over and over again, that DNA can reveal the exact and true identity of people, past and present. How is this justified?

We could talk about it in terms of a "truth effect"—an illusory trick of the mind based on our cognitive tendency to accept a statement as true if we have heard it enough times.[3] Thus, we tend to accept the superiority of DNA as a means to find complete answers to prehistoric people's identities—who they were, exactly—although we know how little DNA can actually grasp of all the wonderful richness that is human life, if we only take a moment to read and reflect.

This Is DNA

In a material sense, DNA is a molecule. In the nuclei of most cells in a human body there are long strands of DNA molecules, which are called chromosomes. The chromosomes contain genes. Genes are sections of DNA that are involved in coding the production of protein, enabling the organism to be built and maintained through its lifetime. The genes thereby influence the forms and physical functions of the body. But the chromosomes are more than genes. For the most part, around 98 percent, they consist of noncoding DNA. Some of this serves functions for the cell, but quite a large portion of the noncoding DNA has as-yet-unknown functions, and until very recently was believed to be of no use at all, a kind of "dark matter" passed down to us from our ancestors.[4] Together the chromosomes make up the genome—the whole set of DNA for an individual. The genome contains chromosomes with coding and noncoding DNA in the cell nucleus, plus a small separate chromosome found in the mitochondrion, an organelle that functions as the cell's energy generator.

At conception, the human embryo receives half of its nuclear DNA from the sperm and half from the egg. That also means that half of the DNA from each parent is lost, and cannot be traced in the new individual. Which sections are inherited from the maternal and the paternal sides are random, as they are recombined before they become the new embryo's chromosomes. But one small part of the nuclear DNA does not recombine: the majority of the Y chromosome that is only found in male individuals, and is thus inherited intact from father to son. Similarly, the mitochondrial DNA is inherited intact, without recombination, from mother to child. So if your nuclear DNA is a more or less wild mix of recombined genes (2 percent) and noncoding DNA (98 percent) from your birth parents, your mitochondrial DNA will always be the same as that of your mother, and, if you are male, your Y-chromosome DNA will be exactly the same as your father's—as long as there has not been a mutation, a slight shift in the order of base molecules, during the copying process. Mutations, which occur in the nuclear, mitochondrial, and Y-chromosome DNA, are crucial for studies of DNA because they result in a slightly different sequence of base molecules that is passed on to the next generation.

The Episteme of DNA

If we want to talk about, study, and make use of this material that we call DNA, we must create an apparatus of knowledge around it. In this knowledge apparatus—the episteme of DNA—the material is classified (as a molecule), is given a name (DNA), and is associated with certain functions (inheritance, reproduction, and production of protein).

The episteme of DNA has been built since 1869, when it was first detected and described as a separate chemical substance. It was named nucleic acid (NA) in 1889; the deoxiribose component (D) was added to the name in 1919; and the double helix shape of the molecule was presented in 1953.[5]

Throughout the nineteenth and early twentieth centuries, there was no clear idea about the function of the DNA substance, and it was not until 1944 that its role in inheritance and reproduction was established with the connection between the substance and the genes. The word *gene* was first coined in German in 1909 and meant something like "element of exact heredity," but it had at that point no connection with the obscure DNA substance. In 1944, the two were brought together in one and the same episteme, when DNA (then associated with chemistry) was connected to ideas of heredity based on Mendelian inheritance (principles of heredity established by Gregor Mendel, a Catholic monk experimenting with peas in the mid-nineteenth century). With this connection, DNA was identified as a key to understand the reproduction of life. Hence, from 1944 onward, the episteme of DNA became intimately connected to genetics and biology—the science of life—in the new field of molecular biology. Ever since, this field has investigated the functions of DNA in organic life. It has contributed to knowledge that is fundamental for medical research and has enabled a broad range of important technologies, from cancer diagnosis to the development of vaccines.

Over the course of the twentieth century, the study of human DNA expanded into several distinct scientific fields. In molecular anthropology, DNA studies gave new insights into questions of human evolution. In population genetics, which is a branch of evolutionary biology hinging on mathematics and statistical modeling (which existed already in the 1920s, before DNA came into the picture), DNA was used to investigate ancestry and disease patterns by charting genetic relations between groups of humans.[6] And in forensic science, the development of DNA profiling in the 1980s brought a focus on individual human DNA, which has since been used, for example, in paternity tests and "genetic fingerprinting" for forensic investigation. These three branches are still active today, and are also fundamental to the study of ancient DNA.

The Images of DNA

If we truly want to make sense of DNA and the role it plays in our lives and societies, it is not enough to think about it as a molecule and material studied within science. Already in 1953, the first visualization of DNA as a double helix (fig. 1) was welcomed by a popular media frenzy that forever celebrated its creators James Watson and Francis Crick at the expense of their colleague Rosalind Franklin, who took the first photograph indicating the helix shape. In tandem with a general postwar enthusiasm for science and technology to forward the positive development of humanity, there was growing popular interest in DNA as an objectively detectable "code" or "blueprint" of life.[7] In this wave of enthusiasm surrounding genetic science, many also discovered that DNA could be a lucrative business. Over the second half of the twentieth century, the episteme of DNA became intimately entwined with corporate business, circling around the market value of patents and ownership of DNA data, and geneticists were awarded glory and fame with an array of Nobel Prizes.

Compared to other fields of scientific inquiry, such as geology or gerontology, genetic science has had an extraordinary popular breakthrough. And as pointed out by media scholar José van Dijck,

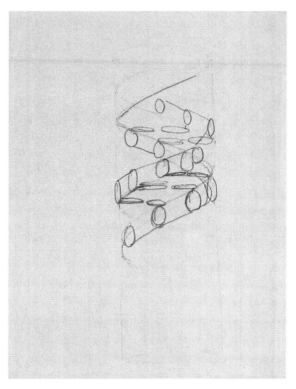

Figure 1. Pencil sketch of the DNA double helix by Francis Crick. Courtesy of the Wellcome Collection, CC BY 4.0 DEED.

the formation of genetic knowledge "is not uniquely contingent on the advancement of science and technology, but is equally dependent on the development of images and imaginations."[8] From Crick and Watson's double helix to the suggestive blueprint metaphor and on to the map and the computer metaphors that are common today, images and imaginations have played a crucial role in generating value for genetic science and its related businesses. These images and imaginations have become integral to genetic science to the degree that it is not possible to make sense of DNA without them. When we talk about the episteme of DNA, we must thus embrace both these aspects—DNA as a molecule and material, *and* DNA as image and imagination.[9]

This Is aDNA

Ancient DNA—aDNA for short—is the DNA from ancient organic remains. This book will focus on aDNA from human remains, but it is also possible to study aDNA from animals, plants, pathogens, and other organisms. Chromosomes of aDNA are always more or less decayed, since the DNA molecules stop replicating and start to fragment once the organism has died. They continue to deteriorate with time, particularly under hot and dry conditions. In dry bones, therefore, the DNA is more deteriorated than in remains of hair and soft tissue, found, for example, in permafrost mummies and bog bodies.

If you want to study DNA (whether from an ancient or a living organism), you first have to collect cells containing DNA from that individual. In living humans, you could, for example, take a sample of blood, semen, or cheek cells. In ancient human remains, the sample is usually taken from bones or teeth, and in rare cases from preserved hair or soft tissue. Then you need to release the DNA by breaking up the cells. This can be done with different methods, which put the bone powder or cheek cells through physical and chemical treatments that isolate the DNA molecules from the other components of the cell.[10] In the decayed DNA from ancient remains there remain only small fragments of the original genome in the cells, so you need to "amplify" the volume of these fragments to be able to work with them. The amplification is often done with the method of polymerase chain reaction (PCR), a kind of "molecular photocopying" in which the genetic material is replicated to create a greater volume of available fragments.[11] It is done by exposing the DNA fragments to heat and enzyme, often in a "thermocycler" machine. Once amplified to a larger volume, the DNA can be analyzed by means of sequencing, which determines the order of the bases, the four chemical building blocks in the DNA molecule: adenine (A), cytosine (C), guanine (G), and thymine (T). The sequencing can be done by hand, but nowadays it is performed by powerful computers. The resulting "code" showing the sequence of the bases (e.g., ACGTTTGC-CCAGAGA) is the geneticist's key to the function of a particular DNA segment. Some sections of DNA will contain genes, while others may have regulatory instructions that switch genes on or off.

The amount of DNA that can be extracted from a given sample varies depending on the volume of the sample and the degree of preservation, which is in turn affected by the density of the sampled tissue. The petrous bone of the inner ear, for example, has been identified as particularly dense, and therefore often contains DNA even in ancient human remains. In soft tissue (as might be found in remains that have been preserved in ice or in a bog) the chances of extracting more DNA for analysis are greater, and so a smaller sample would be needed. Importantly, however, since aDNA is always more or less deteriorated, the aDNA analyses you will read about in this book cover far from the entire genome, which contains 6 billion base molecules. Most current studies of aDNA are built on analyses done (with confidence) on only a small proportion of the genome, equivalent to perhaps a few percent, and in many cases even less.

In the early days of genetic science, it seemed impossible to work with aDNA because of deterioration. In early DNA research it was difficult enough to extract and sequence DNA from living organisms, and most of the early studies used DNA from organisms with much smaller genomes than the human, such as fruit flies. Hence, for a long time it was assumed that deteriorated aDNA from humans and other mammals with genomes containing billions of base molecules would never be subject to successful scientific investigation. For ancient human DNA, it was also a question of avoiding contamination from the abundance of very similar DNA from living humans that was swirling around archaeological excavations, museums, and laboratories.

Yet some geneticists experimented with aDNA from animals, and in 1984 a team in a lab in Berkeley, California, managed to extract and sequence DNA from a quagga—a zebra-like mammal known to have been extinct since the nineteenth century—using a sample of dried muscle tissue from a museum specimen.[12] If the DNA sequence itself was perhaps of less analytical interest for the study of ancient humans, the successful extraction and sequencing proved that it was methodologically possible to work with mammal aDNA.

Svante Pääbo—then a graduate student, now a Nobel laureate and global figurehead for aDNA research—worked in the following years on mummified human remains from Egypt and Florida, from which he eventually managed to extract and analyze mitochondrial DNA in 1988.[13] This was a breakthrough in the sense that it was the first successful analysis of aDNA from ancient human remains, but it was only the small, separate mitochondrial DNA, and only in the very rare cases where soft tissue had been preserved—such as mummies in deserts or permafrost—was it possible to work with ancient human DNA. For the more common human remains in the form of dry bone from archaeological excavations, the challenges of deterioration remained overwhelming.

When most researchers had surrendered and deemed it impossible to work with ancient human DNA from dry bone, Oxford geneticist Erika Hagelberg found a clever solution to the problem of contamination and set out to extract DNA from ancient pig bones. After a year of experimentation with the recently developed PCR method, she managed in 1989 to sequence aDNA from dry bone—evidently uncontaminated, since there were no other pigs in the lab.[14] This result broke the spell. Even though the problem of contamination remained overwhelming for human aDNA, Hagelberg had demonstrated that it was possible to extract and sequence aDNA from dry mammal bones. With that, DNA fully entered the realm of archaeological science as a potential means of investigation into the ancient past.

The first successful extractions and analyses of ancient DNA in the late 1980s were no doubt great methodological achievements enabled by assiduous laboratory work and facilitated by new technological solutions like PCR.[15] But it should be noted that these developments did not generate any new structures of knowledge. In other words, the study of aDNA added a new study material, but did not entail any epistemological reform. The nascent field of archaeogenetics instead conformed to the epistemology of preexisting branches of human genetics from which it took its three principal lines of inquiry: evolution (primarily from molecular anthropology), ancestry (from population genetics), and individual profiling (from forensic science). An important early aDNA study along the evolution line, for example, analyzed a mitochondrial DNA sequence from the remains of a Neanderthal individual and concluded that Neanderthals disappeared from the evolutionary history of humanity without interbreeding with modern humans.[16] A study of ancestry investigated genetic relations between ancient populations in the Pacific Islands and North America.[17] And individual DNA profiling was used in studies on archaeological celebrities such as Ötzi, the prehistoric "ice man" from the Tyrolean Alps.[18] These early studies were fragile because of problems of deterioration and contamination, and there was quite a fierce debate about quality and methodological rigor among scientists in the field at the time.[19] Indeed, some of the conclusions of the 1990s (including that of the relations between Neanderthals and modern humans) were later withdrawn and revised.[20] Scientifically rigorous or not, however, the public interest was intense. Studies of aDNA were presented in the most prestigious journals, and the results made newspaper headlines.[21]

Along with the technological developments and intense public interest in ancient DNA came the first critical inquiries into the ethical aspects of this new and growing field of research. Scholars and scientists in Europe and the United States urged researchers of ancient DNA to be sensitive to the ethics of dealing with human remains for purposes of scientific research. Most of all, it was important to recognize that the "samples" of bone and tissue employed for scientific DNA analyses had human and nonhuman contingencies with social, emotional, political, and legal consequences. In other words, ancient things and the remains of ancient people tend to matter to living people, beyond the realm of scientific inquiry. In this vein, anthropologists Frederika Kaestle and K. Ann Horsburgh wrote that the ethical issues of ancient DNA research pertain to the destructive analysis of bone and other physical remains as well as the "much more complex" issue of accountability, which includes "the responsibility of the researchers to consult with groups that may be affected by the research." In the contexts of living communities, they write, "the results of aDNA studies may impact the social, political, and legal situation that living groups find themselves in, and may contradict or offend beliefs about their ancestors and origins."[22]

The ethical concerns that were raised by Kaestle and Horsburgh more than twenty years ago remain just as crucial today, particularly with regard to Indigenous peoples. Ethical discussions around ancient DNA research are still focused on the destruction of physical remains, responsibilities toward living descendants and stakeholders, and cultural sensitivities surrounding sites, ancestors, and spirituality.[23] We shall return to these discussions further on in the book.

For the broader public, the rising possibilities of aDNA research were exciting, even as they recalled historical examples of the darker side of genetic science. Concurrent with the first wave of comprehensive aDNA studies was a more general excitement around genetic science with the birth of Dolly, the cloned sheep, in July 1996 (fig. 2). A year later was the release of *DNA*, a science-fiction horror movie featuring an unintentional cloning and subsequent revival of an ancient monster in the Bornean jungle. Together, Dolly and *DNA* reflect the two-sided nature of genetic science in the public imagination. While there was hope and excitement about the potential of genetics to create a better future by manipulating and copying DNA, there was at the same time a fear of the consequences of "messing with Mother Nature," as the tagline read on posters promoting the film. The fear also fell back on the intimate interlacing of the history of genetics with the politics and practices of eugenics, and its historical association with genocide in the race experiments of the Nazi regime in the 1940s. Genetic science has always had to walk a fine line, on a steady mission to convince the public to hope and not to fear.[24]

Even before the *DNA* movie, fantasies of activating DNA to bring the long dead back to life had been used to spice up adventure fiction and archaeo-horror films like *Jurassic Park* and *The Mummy*. And, as

Figure 2. Dolly with Professor Sir Ian Wilmut, leader of the research that led to her birth. Photo courtesy of The Roslin Institute, The University of Edinburgh, Roslin, Scotland, UK.

noted by the sociologists Dorothy Nelkin and M. Susan Lindee, "Popular culture matters. For many consumers, media stories, soap operas, advice books, advertising images, and other vehicles of popular culture are a crucial source of guidance and information."[25] While most people are able to separate scientific facts from adventure fiction, both tend to influence popular knowledge. And aDNA has a unique potential to attract attention as it unites two icons of popular culture: archaeology, with its Indiana Jones allure of adventure and mystique, and DNA, as a metonym of hard evidence and a blueprint of life.

The academic and scientific realities of both archaeology and genetics were, of course, far from these fantastic renderings. In reality, the potential of using human aDNA in early archaeological investigations was limited to a few distinct lines of inquiry, and only in human remains with well-preserved tissue, such as mummies. By comparing the DNA—both genes and noncoding DNA—of several such well-preserved individuals, it was possible to estimate how close or distant their biological relations were (for example, sibling or third cousin). In the genes of one individual, it would also be possible to trace with accuracy some hereditary conditions, such as Huntington's disease, and physical characteristics like waxy ears. These particular conditions and characteristics have a one-to-one relation between genotype (the DNA sequence) and phenotype (the actual observable characteristics), meaning that a certain sequence of base nucleotides (combinations of A, C, G, and T) on a specific gene will always have the same effect on the developed organism. But the genotype is in general far from the same as the phenotype. The absolute majority of our observable physical characteristics and hereditary diseases are, as far as we know, decided in the roulette of life through a complicated combination of several genes, environment, life choices, and pure chance. For the most part, DNA can thus only indicate likelihood of a certain physical characteristic, or disease, and cannot determine with certainty what the individual actually looked like or whether they developed that disease. In other words, DNA is far from a one-to-one blueprint of life.[26] Add to that all the social and cultural aspects that determine how we define our ancestry and identity, and you begin to see how limited the DNA window is, if you want to gain meaningful knowledge about ancient human life.

Yet the general trust in aDNA has by far exceeded this slim window. In the history of aDNA research, adventure fiction such as *Jurassic Park* has captured the interest of the general public and stimulated a notion of aDNA as a complete set of information about prehistoric life, ready to be revived by the next scientific discovery.[27] Much of the hype was driven by popular media, but some geneticists played along. A memorable example is this passage by Bryan Sykes, "an Oxford scientist with a flair for the dramatic,"[28] from his bestselling book, *The Seven Daughters of Eve*:

> Our DNA does not fade like an ancient parchment; it does not rust in the ground like the sword of a warrior long dead. It is not eroded by wind or rain, nor reduced to ruin by fire and earthquake. It is the traveler from an antique land who lives within us all.[29]

From a purely scientific point of view, this is far from correct. Ancient DNA is indeed both "eroded" and "reduced," and there are no travelers living in our own DNA—only scrambled bits and pieces from millions of ancestors that are very difficult, often practically impossible, to trace in any meaningful way back to specific individuals in ancient times. But such images and imaginations emerged and were quickly established in popular discourse. While most lab scientists continued to struggle with painstaking manual sequencing of deteriorated and potentially contaminated aDNA, their experiments attracted media interest, and a general belief in the power of aDNA was maintained by adventure fiction like *Jurassic Park* and encouraged by popular science celebrities like Bryan Sykes.[30] If we want to fully grasp the potentials of ancient DNA, we must, therefore, embrace both aspects—aDNA as a molecule and material, *and* aDNA as image and imagination.

Enter Genomics

Approaching the turn of the millennium, computer-based methodologies enabled new forms of genetic research on living organisms. The widely acclaimed Human Genome Project (HGP), which had been bolstered by a veritable "geno-hype" to secure public support and funding,[31] and which US president Bill Clinton and UK prime minister Tony Blair jointly announced a success in the year 2000, made it possible for the first time to document all parts of the human DNA—the entire genome of a human being.[32] Clinton turned to metaphors of global colonization and likened the record of the human genome to a map from "courageous" colonial-era expeditions around the world, describing it as "the most important, most wondrous map ever produced by humankind." And, added Blair, it demonstrated "the way technology and science are driving us—fast-forwarding us—all into the future."[33] Cushioning the bombastic techno-colonial message was the project's antiracist ethos. Demonstrating that any given human genome is 99.9 percent identical to that of other humans,[34] it ostensibly provided scientific evidence that human races do not exist. The applause was deafening and the popular media breakthrough immense.

In the history of genetic science, the Human Genome Project marked the turn to an era of genomics and bioinformatics, in which computers did much of the groundwork and genetic analyses were based primarily on statistical modeling and big data. A new set of methods, commonly known as "high-throughput sequencing," allowed computers to do most of the laborious work of sequencing, to determine the order of the nucleotide base molecules (A, G, C, and T) after the DNA extraction, which had previously been done mostly by hand. With the new sequencing methods, a computer can work through the full length of a human genome—around 6 billion base molecules forming 3 billion base pairs—in a day, compared to the year required by previous methods. Genome-wide analyses could be done faster, and at less cost, than ever before. With the new methodology it was suddenly possible to study larger sections of an individual genome, and there were aspiring hopes for a new kind of individual-focused and custom-made genetic science. The mapping and analysis of every person's unique genetic composition would then replace and make redundant the earlier categorization of genetic profiles in terms of ethnicity or race. So it was believed at the time. Further studies, however, proved the functions of the genome to be much more complex, and not easily mapped on an individual scale. Hence, the categorization of genetic profiles in terms of ethnicity or race has not waned or disappeared in the wake of the HGP, quite the contrary.[35]

Let us stop for a moment and say a few words about terminology. The term "genome-wide analysis" is sometimes used synonymously with "full genome," but it is not in fact the case that they analyze the entire genome of 3 billion base pairs. It is an indication of the complexity of analyzing an entire human genome that only 92 percent had been mapped when Clinton and Blair announced the successful completion of the HGP in 2000, and the rest was completed as late as 2022.[36] This was indeed an important breakthrough, but it is worth noting that the aim was to generate a *composite* genome, constructed from several different individuals. At the time of writing, the first sequencing of an *entire* genome from one single living individual was still waiting to be completed.[37] This is how complicated it is. So, if we are more transparent with the terminology, a "full-genome" or "genome-wide" analysis targets *specific small sections* of genes and noncoding DNA that are found in different locations across the whole width of the genome.[38]

For population genetics, which had always hinged on mathematics and statistics, the turn to bioinformatics also facilitated work with big data and mapping of genetic variations between groups in global-scale endeavors, such as the Human Genome Diversity Project and the Genographic Project. But the entrance of bioinformatics was a game-changer for human genetic research overall. It created new opportunities for medical research and inspired a plethora of private companies to offer genetic tests claiming to identify their customers' real skin types, indicate their ideal food intakes, or reveal their genetic ancestry in the form of ethnic pie charts. Again, many found DNA to be a lucrative business.

The growth of such businesses relied on customers' belief in DNA as a complete blueprint and accessible book of life—images that are not sustainable from a scientific point of view.[39] For the successful businesses and their associated scientists, however, it made all the sense in the world to fuel rather than challenge these beliefs. One company, for example, promises customers that genetic ancestry tests will allow them to "travel back in time to gain a clearer picture of where you came from, where your ancestors lived and when they lived there."[40] The images that could very well pop up in your head reading this—the prospect of meeting your real ancestors face to face and learning about their personal history, in the actual places where they lived—are far removed from what the results from such a test can actually say and do. But test kits are sold on hopes and aspirations, and in many parts of the world DNA testing has become something of a mass movement. With the broad uptake of these products and tantalizing marketing messages delivered by actors in white coats, inflated beliefs in the powers of DNA spread beyond popular science to the broader realm of commercials and consumption.[41]

The "aDNA Revolution"

The turn to bioinformatics and the development of high-throughput sequencing (also referred to as second-generation or next-generation sequencing) had a significant impact on aDNA research, too. As in other forms of genetic science, the new methods made possible quicker and less costly analyses of larger volumes of DNA. The computerized sequencing methods enabled analyses of deteriorated aDNA, with parallel sequencing of thousands of copies of the same genome, where computers did the jigsaw by comparing copies and putting the pieces and fragments together in the right order. But it did not overcome the problem of deterioration and decay—often it is as little as 0.2 percent of the original genome that remains in samples of aDNA—and the substantial 99.8 percent gaps cannot be filled in other than by qualified guesswork and comparison with other, more complete genomes that are thought to be similar.[42] But the new high-throughput sequencing methods helped to amplify the existing remains, and thereby maximized the possibilities of "reading the code" of those rare preserved sections. The new sequencing regime also made it easier to differentiate the fragmented aDNA from modern DNA of people who had handled the sample. Altogether, this meant that analyses of DNA from dry ancient bone could now be done at a larger scale, with much less risk of having the analysis ruined by contamination.

With such opportunities rising on the horizon, a number of laboratories scaled up their work on aDNA. The field of archaeogenetics saw a marked increase with the first important post–high-throughput-sequencing papers published around 2009,[43] followed by a wave of published results from 2015 onward. Several laboratories in Asia, Europe, and North America were visible in the forefront of these methodological developments, which soon began to be described as a "revolution." Already in 2014, the Scandinavian archaeologist Kristian Kristiansen wrote about a "third science revolution in archaeology,"[44] and the American geneticist David Reich continued the campaign with his bestselling book, *Who We Are and How We Got Here*, in which he refers to "the revolution" eighteen times in the introduction alone.[45] Science journalists reported on "the revolution" as if it was an indisputable fact, often with metaphors of war and violent destruction. A top journalist in *Nature* described the state of the research as "a battle for common ground,"[46] and another in *Science* wrote about the "revolution in human evolution: as it smashes disciplinary boundaries, ancient DNA is rewriting much of human history."[47] It was a powerful and suggestive message that soon reached a broad public. "The revolution" has since become an oft-cited point of reference in the field.

But what, exactly, happened in this revolution? My colleague Daniel Strand did a trawl and close reading of all published mentions of the "aDNA revolution" but found no clarifying explanations, only statements suggesting it had taken place.[48] This indicates that the "aDNA revolution" cannot be regarded as a "scientific revolution" in the classic sense, as defined by Thomas Kuhn, when a scientific discipline is forced to overthrow its previous structure of knowledge and establish an entirely new paradigm in the face of an anomaly—a profound problem that undermines the foundations of the discipline.[49] Whatever happened in aDNA research with the entrance of high-throughput sequencing and bioinformatics, it did not generate any fundamentally new structure of knowledge in archaeology, or in genetics, for that matter. In archaeology, aDNA studies have upheld the teleological Stone, Bronze, and Iron Age structure of knowledge that has been fundamental to the discipline since the mid-nineteenth century. If anything, it has been argued, aDNA studies have revived outdated ideas of human identity and migration from the early twentieth century.[50] For genetics, the knowledge structure of aDNA research still conforms to the three branches of human genetics that have always dominated historical studies involving DNA: evolution, ancestry, and individual profiling. And in terms of research methodology, the new genomic opportunities mostly brought embellishments and upscalings of fundamental research structures that had been developed and established decades earlier.[51]

Rhetoric and Storytelling

So what, then, is the "aDNA revolution"? First and foremost, it appears as a figure of rhetoric, a sign of the indisputable importance of developments in the field. Almost always mentioned alongside is a steep increase in numbers—numbers of analyzed samples, sampled individuals, sequenced genomes in aDNA databases, and published papers. In this vein, Harvard geneticist David Reich has described the "aDNA revolution" as "an avalanche of discoveries based on data taken from the whole genome."[52] The fixation on numbers reflects the turn to bioinformatics and big data, a significant methodological development that can indeed benefit some archaeological lines of inquiry. But again, this is hardly a revolution of archaeological knowledge.[53]

From the perspective of archaeology—the field that has most commonly been connected with the new genomic science in the rhetoric of the "aDNA revolution"—we can compare the turn to big data to a computerized analysis of flint axe measurements. Such analysis can offer insights into relations between flint axe shapes, and more flint axe samples will give more data points for comparison. But this is never enough if we want to know more about the meaning of flint axes in ancient societies. Thus, the archaeological pursuit of knowledge is largely dependent on qualitative analysis and interpretation. Context is key: a flint axe is likely to have been endowed with different meanings in an ancient society if it was found in a grave, in a cache, or on a workshop floor. And the delicate inquiry into context can never be replaced by quantitative analyses of detached samples, no matter how big the data.

Archaeology is therefore also to a great extent a storytelling practice. It would not be possible to make sense of ancient things, or of molecules from ancient people, unless we tell stories about them. Stories are the meaning-making glue that connects disparate fragments and allows us to make sense of them. Archaeological stories contain not only words, but also structured representations of numbers, dates, maps, objects, and images. Endowed with the authority of science and academia, archaeological stories can establish firm ideas about who was present, who was first, and who was important in times when there were no or few written records. Serious archaeological storytelling is far from careless speculation or fabulation. It is anchored in available facts, and it assumes responsibility for each of the steps it takes to turn material fragments into meaningful narratives.

Nowadays, archaeologists are trained to appreciate the social and political power of their stories. They know that their stories are volatile—exciting and empowering for some, but agents of social vulnerability and exclusion for others. From historical examples, we know that archaeological stories can help to build foundations for new nations, just as they can serve to justify war and genocide. In recent years, for example, a despot in a country not far from where I sit has used a self-composed story about the historical origin of his nation to justify atrocious, unprovoked attacks on a neighboring country.[54] In such a context, archaeological stories speaking of ethnic groups in prehistoric times can be as explosive as Molotov cocktails. Archaeology is far from a risk-free enterprise, and it must be performed with caution.

Mindful of these predicaments, it might be tempting to think of the "aDNA revolution" as a symptom of hubris among enthusiastic geneticists and journalists who are ignorant of the realities of history-writing and the volatile political potency of archaeological storytelling. But inflated statements about the revolutionary power of aDNA have not only been made by geneticists and science journalists. A prominent Scandinavian archaeologist, for example, has described aDNA as "a new door to previously hidden absolute knowledge" that will "reduce the amount of qualified guessing" and reveal human history "without having to resort to storytelling."[55] This, of course, is not true. DNA is a molecule, and molecules do not speak. Someone has to tell the story.[56]

And if you look at it, there has been no diminution of archaeological storytelling in tandem with the alleged aDNA revolution. On the contrary, the twenty-first-century methodological developments around aDNA seem to have radically pumped up the volume on the stories being told. The bigger the data, the bolder the stories. As we will see in the following chapters, stories of findings and discoveries in aDNA research have tended to range far beyond what the results of DNA analysis can tell us. In many of the cases featured in this book, the primary role played by aDNA in stories about the ancient past is metaphorical rather than analytical, offering a seal of scientific proof to existing narratives, enticing fictions, and suggestive political agendas.

The God Trick

You may have sensed that I am concerned with the rhetoric of the "aDNA revolution." I am. But let me clarify my concern. In a classic essay from 1988, the American biologist and historian of science Donna Haraway makes a plea for *situated knowledge*—a form of responsible objectivity in science that stands firm against claims of totality and closure, just as it resists relativism. It stays true to science as an enterprise concerned with facts and reality by acknowledging that any knowledge claim must be made from a body in movement—a situated body with a certain previous experience, and an ambition to go somewhere.[57] The claim is as brilliant as it is simple. We know that molecules have no voice, so someone else has to speak about them. And that speech inevitably has to be uttered from a body in movement. All responsible scientific knowledge must therefore fall back on scientists' imperfect positions and partial perspectives. If these situated positions are not transparently accounted for, there is reason for vigilance.

The first enemy of situated knowledge is "the god trick of seeing everything from nowhere,"[58] often by reference to impenetrable technological apparatuses that will replace thinking and make critical scrutiny unnecessary because of the neutral totality of their vision. As far as I can see, the "aDNA revolution" stands out as an almost perfect god trick. In a research field already obscure to most people outside the genetics labs, the turn to bioinformatics thickened the walls of the black box,[59] obstructing critical insight into the actual DNA analyses. Indeed, in most popular communication following the "aDNA revolution," big data and genome-wide analyses have come to indicate completeness and neutrality. Results have been presented as complete and neutral representations revealed by unbiased computers and generated by the body of data itself.[60] But this is not a correct image of genetic science. All geneticists know that a genome-wide aDNA analysis does not come close to covering the entire genome. Often there are only small parts available because of the molecules' degradation. But even in a well-preserved sample, someone makes the choice of which sections to include in the analysis. For the statistical modeling, which is a crucial part of DNA analysis, someone has to choose the variables, models, and parameters that are fed into the computer. And computers are not neutral machines; they have to be programmed by situated bodies. In other words, a genetic analysis is a creative process rather than a neutral rendering of existing information. Thus, all genetic analyses depend on series of choices made by the researchers, and limitations set by the methodological frameworks and traditions in which they work. Furthermore, scientists working with aDNA need to make meaning out of what is basically a set of molecules. Archaeogeneticists have to put words to a material that is essentially wordless. Someone has to tell the story.

My concern, therefore, is with grandiose stories about people and societies in the ancient past that have gained credence with the god trick of aDNA genomics. Such stories point to aDNA as a petrified book of life, a historical source that has neutrally and objectively been cracked open by genetic science and will always trump existing knowledge about prehistoric people because of its evidential superiority. They hinge upon the illusory truth effect of DNA as complete and superior evidence that existed long before high-throughput sequencing, but they have had an extraordinary academic and popular breakthrough with leverage of the revolutionary rhetoric.[61]

Like the vast majority of my fellow archaeologists, I am not negative or skeptical toward the DNA molecule as such. On the contrary, I am convinced that it plays an important role in our lives because of its crucial influence on our physical beings and corporeal functions. I am also convinced that aDNA from humans and other life forms—from mammals to plants and bacteria—can be used in scientific analyses to the benefit of our knowledge of the ancient past. This may seem trivial but deserves to be pointed out, since much of the earlier criticism of grand stories of aDNA has been caricatured and dismissed by proponents of the field as being unduly negative toward uses of DNA in general, and critics have been portrayed as anxious and obstructing of scientific progress.

In an essay in the *New York Times Magazine*, journalist Gideon Lewis-Kraus describes the debates between the proponents of the "aDNA revolution" and critical observers as "a division between two alternate intellectual attitudes: those bewitched by grand historical narratives, who believe that there is something both detailed and definitive to say about the very largest questions, and those who wearily warn that such adventures rarely end well."[62] This book obviously departs from the latter position, but I have a broader aim than to issue weary warnings. In the following chapters we will scrutinize the makings and foundations of grand historical narratives based on aDNA research and ask where the stories come from, now that we have established that DNA molecules have no voice of their own. And since molecules do not speak, this book is offered in defense of DNA. Molecules should not be burdened with responsibility for stories they have not told.

Unlearning the "aDNA Revolution"

That said, there are good reasons to issue warnings, for god trick–induced stories speaking of aDNA can be dangerous. In the past decade, politically potent and socially sensitive stories about the prehistoric existence or nonexistence of human races, claims of the presence or absence of certain ethnic groups in a certain territory, and vivid accounts of professional, social, and gendered identities of individuals in ancient times have been presented as unbiased discoveries in machine-processed data. Being told as total and neutral visions sprung directly out of molecules, these stories are often presented as the final word or closure of a long-standing issue or debate in archaeology. As such, they tend to obscure complexities and naturalize power-claims that are always and necessarily part of archaeological history-writing. This is not to say that all stories of aDNA are harmful, but that they are potentially dangerous when their claims are total and their ambitions concealed.

The following chapters will explore notable stories about aDNA, scrutinizing the space between molecule and narrative with the hope of clarifying what the molecules can and cannot do, as we set out on the slippery venture of telling stories about the ancient past. I will argue that we need to unlearn the evangelical rhetoric of the "aDNA revolution" and the god trick it encapsulates. While we acknowledge the real and potential benefits of using genetic science to inform historic scholarship—of which there are plenty—we should also declare the limitations of these methods and make sure that we have enough knowledge of the complexities of both genomics and historical research to steer clear of pitfalls with dangerous political and ethical consequences. Hence, there is good reason for all parties involved in storytelling with aDNA to listen and learn more from one another. This means also that we need to sit with, and dig deeper into, inter-disciplinary conflict, rather than attempt to disguise or overrule it. DNA molecules have no bad intentions. But when we exploit them for purposes of history-writing we need to honor ambiguity and resist the allure of total vision,[63] for human intentions and identities are always complicated, at times contradictory, and can only be described from situated positions.

FURTHER READING

De Groot, Jerome. *Double Helix History: Genetics and the Past*. New York: Routledge, 2022.

Jones, Elizabeth. *Ancient DNA: The Making of a Celebrity Science*. New Haven: Yale University Press, 2022.

Matisoo-Smith, Elizabeth, and K. Ann Horsburg. *DNA for Archaeologists*. Walnut Creek, CA: Left Coast Press, 2012.

Mukherjee, Siddhartha. *The Gene: An Intimate History*. London: Bodley Head, 2016.

Nelkin, Dorothy, and M. Susan Lindee. *The DNA Mystique: The Gene as a Cultural Icon*. Ann Arbor: University of Michigan Press, 2004.

Reich, David. *Who We Are and How We Got Here: Ancient DNA and the New Science of the Human Past*. New York: Oxford University Press, 2018.

2: Return
of the Arrows

It is often said that we live in times of migration. On the morning news, I heard UNHCR, the United Nations refugee agency, reporting an estimate of 100 million people currently on the move from their homelands, most of them forcibly displaced in acute response to war and violence.[1] The 2021 Nobel Prize in Literature was awarded to the Tanzanian British author and professor Abdulrazak Gurnah "for his uncompromising and compassionate penetration of the effects of colonialism and the fate of the refugee in the gulf between cultures and continents."[2] Political leaders across the world circle around the relations between migration and nation, polarizing political landscapes and forging international alliances on their stances for or against immigration.

It is no surprise, then, that many of our recent stories about the ancient past are also spun around migration. We tend to turn to the past to seek firm ground for unsettled issues, to find answers to questions that concern us in the present. That is just as true in the twenty-first century as it was in the early days of the nineteenth and twentieth centuries, when archaeology provided burgeoning nation-states and waning colonial powers with roots in prehistories as glorious as possible.

The Pots-Equal-People Paradigm

When I was a student of archaeology some thirty years ago, I was taught that the arrow was a thing of the past. Well, not of the ancient past, but belonging to a set of outdated archaeological ideas that had been proven not only invalid but also dangerous. In my textbooks, arrows were foremost featured on maps associated with the so-called *culture-historical* school that was in vogue in European archaeology in the early decades of the twentieth century.[3] Culture-historical archaeology put focus on archaeological "cultures," which were defined by distinct material forms and styles in pottery and other things that were produced and used in ancient societies. Characteristic of culture-historical archaeology are illustrations with continental maps, on which arrows indicate historic migrations of cultures from one area to another.

The culture-historical perspective was first voiced and established in Germany in the late nineteenth century, from where it gained traction in archaeological societies around the world. Originally launched as a more detailed and culture-sensitive alternative to evolutionary explanations of prehistoric developments, it soon merged with the intense interest in questions of race and nation that characterized early twentieth-century politics and academia. Names of typical artifacts that archaeologists used to label their cultures, such as bell-beaker pots or Clovis spearheads, were suddenly turned into ethnonyms, as in "Bell-beaker people" or "Clovis communities."

In this pots-equal-people paradigm, ancient societies were looked upon as essential and naturally bounded units endemic to certain territories, much like the then-rising concept of the nation-state. With bounded stability as norm, cultural change was explained by "diffusion" in the form of migrations of ethnic groups from one area of settlement to another—like drops of ink in a glass of clear water. By the same logic, each ethnic group would have a certain essence and creative genius that produced a distinctive "ink" in the form of technology and artistic style. When a different style or technology appeared in the ancient layers of an archaeological site, archaeologists would thus read it as a record of the arrival of a new group of people with a superior strength and genius, who had replaced or interbred with the earlier group to create new or hybrid cultural forms. Consequently, the question of origin was key for the culture-historical paradigm. Where was the ultimate place of origin from which a superior cultural genius had diffused and migrated? This was a question of great political potency. Evidence of a prehistoric grand culture, the presence of a superior genius in your own territory, offered prestige as an anchor to the modern development of the nation-state. Maps with arrows running across and between continents were used to visualize connections between origin and telos, and to explain ancient population movements as agents of cultural change.

The arrow was an effective illustration as it conveyed in one sign a number of things simultaneously: timeless essence (where different-colored arrows marked the distinctive characters of groups on the move), direction (in the movement from one geographical location to another), and temporality (sometimes hundreds or even thousands of years from the tail to the pointed tip).[4] It proved so effective that it was soon established as a standard visual element not only in archaeological research papers, but also in popular archaeology books and educational posters for the classroom (fig. 3).[5]

When I first encountered the arrows in my archaeology textbooks, I had the benefit of historical hindsight. Not only had decades of research in the humanities and social sciences established a much more sophisticated understanding of ethnicity and more complex mechanisms of cultural change in prehistoric times, but also we were able

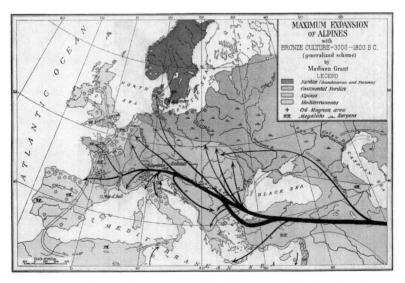

Figure 3. Map from Madison Grant's 1916 book *The Passing of the Great Race.* Reprinted from Madison Grant, "The Passing of the Great Race," *Geographical Review* 2, no. 5 (November 1916): 354–60.

to see the full social and political consequences of an extreme reading of culture-historical archaeology. For as it grew more popular in the early twentieth century, the culture-historical take on prehistory had proven useful for people and political movements advocating racial purity, from the influential American eugenicist Madison Grant[6] to Adolf Hitler's Nazism. Particularly the latter, which used archaeological theories of Aryan peoples' superiority as pretext for genocide, left a shameful burden on the archaeological community in the decades after World War II. It put a stigma on archaeologists associated with the Nazi regime, such as Gustaf Kossinna (who died in 1931, but whose works have been described as the precursor to Nazi archaeology), and relegated much of the culture-historical school and its continent-spanning arrows to the historical garbage dump.[7]

And now they are back. The arrows are everywhere. After a half-century on the garbage heap of tried and rejected archaeological models, they have risen from the ashes as standard visuals in studies of aDNA. With an iconography reminiscent of a game of *Risk*, multicolored arrows span entire continents and reach every corner of the world map. In academic papers and news stories, they are seen as effective illustrations of prehistoric peoples' movements, from the evolution of the first humans in Africa and their descendants' subsequent spread over the globe, to "massive migrations" of Yamnaya herders across Bronze Age Europe.[8]

What are we to make of the resurrection of arrows in archaeogenomics? Do they carry the same messages as the arrows that illustrated Grant's ideas and Kossinna's works, or are their connotations entirely different?[9] What is the political potential of this new generation of arrows in our own era of migration? We shall return to these questions below, but first let us review two recent aDNA studies to see what steps they have taken to get from molecules to arrows.

Two Examples

Our first example is a 2020 study of prehistoric East Asia by Chao Ning and colleagues.[10] It is based on analyses of fifty-five ancient human genomes "from various archaeological sites representative of major archaeological cultures across northern China," dated to between 7,500 and 1,700 years ago. Its focus is the so-called neolithization process, the turn from hunting-gathering to agriculture as a main subsistence economy, which has been a long-standing preoccupation of the archaeology of East Asia as well as of many other parts of the world. Archaeological investigations had already established that there was millet cultivation in both the Yellow River Basin (now in central China) and the Liao River Basin (now in northeast China) around 8,000 years ago. While crop cultivation quickly came to dominate in the Yellow River Basin, archaeological finds suggested that there had been a more fluctuating subsistence economy with periods of nomadic pastoralism in the Liao River Basin. Hence the main question addressed by this new study was whether the observed variations in subsistence economy around the Liao River had been owing to changes in climate or to "substantial human migrations." By making comparisons in time and space between the 55 analyzed ancient genomes and 2,077 genomes of now living "Eurasian" individuals in a reference database, the study indicated that ancient people living in the regions of the Yellow River Basin, the Liao River Basin, and the Amur River Basin to the north belonged to distinct genetic clusters corresponding to their geographical location—with people in the geographically remote Amur River region being more genetically distinct from people in the Yellow River region than were the people in the Liao River region. It also showed that the ancient people in the Yellow River region were genetically similar, if not identical, to present-day Han, which is the dominant ethnic group of the People's Republic of China. Based on the observation that the ten sampled ancient individuals from the Liao River region had genetic profiles that corresponded alternately more with the sampled individuals in the Yellow River region (where crop cultivation was dominant according to archaeological investigations) and those in the Amur River region (where archaeology suggests hunting-gathering and pastoralism were combined with crop cultivation), the study linked fluctuations in crop cultivation and pastoralism around the Liao River to "admixture events," which is a genetic term for when one distinct (often referred to as "isolated") population interbreeds with another. It thus concluded that shifts in subsistence economy in the Liao River region were owing to migrations of people from the Amur and Yellow River regions, respectively.

Our second example is a group of recent studies on the peopling and population history of the Americas, as summarized in a 2021 review article by Eske Willerslev and David Melzer.[11] Altogether, the studies comprise ancient human remains from around 340 individuals, of which 120 were analyzed with whole-genome analyses and 220 with analyses targeting specific sections of the genomes. The most ancient samples date to 12,800 years ago, and the most recent are from between 1000 and 2000 CE, so the studies included 340 individuals who lived at some point in a time period of approximately 12,000 years and whose remains had been found in different sites across North and South America. These samples were compared with analyzed genomes of pre-historic people in Northeast Asia and with those of now-living Native American individuals. According to the article, the results altogether indicate that the first peopling of the Americas was set in motion with a single first migration of "a basal American lineage" from Northeast Asia, which was halted in Beringia by the austere climate and environment of the Last Glacial Maximum around 20,000 years ago. This led to a period of isolation, also referred to as "the Beringian standstill," from which emerged three distinct populations: "Ancient Beringians," "Ancestral Native Americans," and a "genetic ghost population" named UPopA.[12] They crossed into the land we now know as North America when the recession of the ice made it possible for them to pass. The studies indicate that the "Ancient Beringian" population did not go south of the ice sheets, but the "Ancestral Native American" population did, and between 21,000 and 15,000 years ago it split in three lineages: "Northern Native American," "Southern Native American," and one known as "Big Bar." The Northern lineage extended in the north, while the Southern "rapidly spread southward" across North and South America (fig. 4). Judging from patterns of genetic relations, there appear to have been continuous movements of people within the Americas in the following millennia, to and from other regions in Asia, the Arctic, and the Pacific, and later the inflow of people from Europe and Africa that altogether contributed to current patterns of genetic ancestry in the Americas. A key result from these studies, however, is that all genetic evidence points in the direction of Northeast Asia for the first ancestral origins of Native American people. This effectively rules out the already much-contested, but in some fringe circles resilient, idea of a European origin of America's first peoples (an argument based on similarities in the Pleistocene stone technology practiced in the European Solutrean and the American

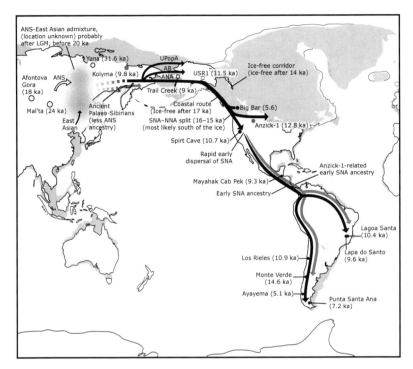

Figure 4. Map of the peopling of the Americas. Adapted from Eske Willerslev and David J. Meltzer, "Peopling of the Americas as Inferred from Ancient Genomics," *Nature* 594 (June 2021): 356–64.

Clovis cultures). Contrary to this outdated idea, the genomic studies conclude that there was a common genetic origin for all prehistoric people of North and South America—and, by association, of Native American people today—in the "basal American lineage" in Beringia some 20,000 years ago.

Samples

Now, there are a few basic things in these studies[13] that need sorting out before we can move on to discuss the stories they have inspired. The first is the number of samples they include. In our American case, the cited studies included whole-genome, or genome-wide,[14] analyses of aDNA from 340 individuals over 12,000 years across the Americas.

That is on average one human being every two generations, buried at a random place on two entire continents. On a similar scale, our first, East Asian case was built on the analysis of fifty-five ancient human genomes "from various archaeological sites [. . .] across northern China" in a time span of 5,800 years.[15]

If you look at these numbers from the perspective of genomic science, where genome-wide analyses of ancient DNA were beyond belief not very long ago, the extraction and analysis of fifty-five ancient genomes is an impressive amount and an amazing achievement. But when you bear in mind the context of prehistoric society, the picture is rather different. If we consider the fifty-five genomes of the East Asia study as belonging to actual, situated human beings, we are talking about a single individual every four generations, in a vast region that was likely home to millions of people at any one time. For an archaeologist attuned to the complexities and local variations of prehistoric culture and society, that is a scant sample, to put it mildly.

If you raise this concern with a population geneticist, they will likely respond that the analysis of a single genome is so much more than a study of a single individual. And that is true. Owing to the fact that we have inherited our DNA from our birth parents, who inherited theirs from their birth parents, and so on, our genomes contain—in theory—traces of all our ancestors back to, and beyond, the origins of humanity in the Horn of Africa some 200,000 years ago.

At the conception of a new individual, there are sometimes hiccups in the process of copying DNA from parent to child: a mutation where the order of base nucleotide molecules is shifted. The resulting variation in the sequence of nucleotides is called an "allele." The simplest form of allele is a "single nucleotide polymorphism" (SNP)—in other words, a base molecule that appears in more than one variant of A, C, G, and T.[16] Alleles such as SNPs are fundamental for genetic analyses. They can affect the physical characteristics and functions of the organism if they are located in the coding part of the genome. But even if they are found in the noncoding majority regions of the genome and will have no visible effects on the living organism, they are inherited and can thus be used to trace ancestry.

A problem, though, is that half of the genome from each parent is lost at the conception of a new individual, and almost all of the 3 billion base pairs contained in a human genome are recombined, so you cannot know which bits came from the father and which came from the mother—with the exception of the Y chromosome and the mitochondrial DNA, which are inherited uncombined from the father and the mother, respectively. Alleles that result from mutations in the mitochondrial or Y chromosome DNA are therefore always and exclusively traceable to the individual's maternal (mitochondrial) or paternal (Y chromosome) birth lineage. These particular alleles are referred to as "haplotypes" (and several haplotypes are grouped into "haplogroups"). This means that from your genome alone, it is practically impossible to trace any of your individual ancestors more than a couple of generations back, if they are not on your straight maternal or (if you are male) paternal line, where the haplotypes and haplogroups are inherited uncombined. And even then, it is not a simple task. So how can you know the genetic ancestry of all people in ancient times from analyses of a few individual genomes across a vast continent over thousands of years?

The simple answer is that you can't. These knowledge-claims are grounded on calculations of probability rather than observation. Statistics and algorithms make it possible to discern patterns of SNPs that exist in the noncoding DNA as well as in the genes, so most of them (around 98 percent) have, as far as we know, no concrete effects on our physical beings. But when they are analytically detected, combined, and compared with SNPs from other genomes, they can indicate patterns of relatedness. In short, if you share most of your SNPs with another individual, you likely share most of your ancestors with them as well.[17] With statistical models using comparative samples from modern humans and chimpanzees, one can calculate degrees of distance between the analyzed samples and thereby simulate probable demographic histories to explain the relations between them.[18] Hence, by comparing masses of genomes, it is possible to construct a virtual jigsaw puzzle of genetic relatedness.[19] Where there are several possible scenarios that could explain the pattern, the strategy of Occam's razor (an approach that seeks the explanation based on the fewest elements possible) is often applied to single out the most plausible one.[20] The result of this choice of strategy in population genetics is that simple models of explanation are privileged over complex ones.[21]

If you only have a few pieces of the puzzle, as is the case in our two examples from East Asia and the Americas, you can use statistical models and algorithms to calculate the most likely—or least unlikely—shapes of the missing pieces. In fact, this kind of modeling and calculation of probability constitutes the bulk of most aDNA studies. In scientific papers, it can read like this:

> F_4 statistics confirm that both groups are intermediate between AR and YR groups, represented by AR_EN and YR_MN, respectively (Supplementary Fig. 18). We adequately model both groups as a mixture of AR and YR groups, with higher AR contribution to HMMH_MN (39.8 ± 5.7% and 75.1 ± 8.9% for WLR_MN and HMMH_MN, respectively; Supplementary Table 4).[22]

For most people interested in prehistoric studies, professional archaeologists included, this is a black box with very thick walls. It is difficult to see the caveats, and mathematics can give an impression of correctness and neutrality. Many of us skip over these parts and go straight to the results. But there is good reason to linger for a while, because some essential building blocks in the stories of aDNA take shape here, in the analyses based on mathematics and statistical models.

Populations

It is said that the genome of a human being is more than 99 percent identical to the genomes of other humans, about 80 percent the same as the genome of a cow, and in no small part identical to that of a banana. Hence it would be perfectly correct to say that any human individual is 100 percent human—just as it is correct to say that every individual is 100 percent unique, because not even identical twins have the exact same genome. Both extremes, however, are useless if you want to say something of interest about people in the ancient past. Imagine a genetic study of an ancient cemetery concluding that all the interred were human beings—or, indeed, that all were unique individuals. There would be no newspaper headlines and no more research grants for those scientists. The reason why we put faith in aDNA must be that we believe it can say something about prehistoric people's identities and relations—beyond the fact that all are human, and all are individuals. Therefore, aDNA studies of human remains always set out to make sense of degrees of genetic difference in the less than 1 percent of the genome where we are *not* all the same.

If we want to discuss genetic difference between people, we need first to decide on a level where we discern the difference, somewhere on the scale from the unique individual to all of humanity. The choice is arbitrary in principle (could be a family, a small town, or a continent with billions of people), but aDNA studies almost always speak of populations, as in "Ancient Beringians" or "UPopA." This is because aDNA studies rely on the methodologies of *population genetics*.

Population genetics started to develop as a field as early as the 1920s, before the function of DNA was known (since genetics existed before the detailed knowledge of the DNA molecule), but saw a considerable expansion with the use of computers beginning in the 1970s. It is now mostly concerned with medical research on living people, often with the aim to find genetic variations that are associated with increased risks of disease. To trace patterns of genetic variation among humans, population geneticists target the parts of the genome that they know differ, and use mathematical calculations and statistical models to discern patterns in the variation of SNPs. To make statistical patterns meaningful and practical to work with, they apply the concept of "population" to designate clusters of genetic proximity.

In modern society, such clusters of genetic proximity tend to correspond statistically to social groups, such as nations or ethnic minority groups, because national borders, official languages, and situations of marginalization have had long-term effects on how people meet and mate. Of course, this is not true for every individual. We can observe among ourselves, our families, and our friends that close genetic relations exist across all sorts of borders. But statistically, people in modern societies are most likely to meet and mate close to home. And likelihood is what matters in population genetics, where statistical patterns are paramount, and simple models of explanation are privileged over complex ones.

By acquiring DNA samples from individuals who fit certain requirements (such as having both pairs of biological grandparents native to the same geographical location, which is a common criterion), and labeling each sample with the reported nationality or ethnic belonging of the person it was taken from, population geneticists construct reference databases of DNA variations. And it is here, in these databases, that ethnic and national genetic identity is born. If you read about "Portuguese" DNA, for example, the term refers to the statistical calculation of the most common genetic variations among people who have been sampled and registered as "Portuguese" in a particular DNA database. It could be 5 individuals, or 5,000, who have been selected by certain criteria and have been labeled "Portuguese." The Dutch anthropologist of science Amade M'charek has called them *convenient conventions*, these words and labels that population geneticists use to facilitate communication during the research process.[23]

Importantly, the convenient conventions of genetic identity in terms of named populations should not be assumed to be the same as social identity in the form of citizenship or self-identified ethnicity—even if they use the same words, which can be quite confusing. For example, if you approach ten Portuguese citizens on the street in Lisbon today, it is quite likely that some of them would not register as genetically "Portuguese," simply because one or more of their biological grandparents were from other parts of the world, and their SNP patterns would therefore not correspond statistically with the people selected for the database, who had all grandparents native to the present-day territory of Portugal. This, of course, does not make these individuals less Portuguese in their civic or ethnic identity, but they will not be identified as genetically "Portuguese" according to the definition of the database.

This creative aspect—that DNA does not contain a detectable ethnic or national code *in itself*, but that the ethnic labels we see in studies of ancient DNA have been *ascribed* by geneticists according to certain criteria applied to samples in a database—is very important to understand when you assess studies based on population genetics.[24] Unfortunately, however, the crucial distinction between "Portuguese" as a convenient convention during a scientific research process and Portuguese as an identity marker of personal, social, and political importance for real people is rarely clarified in popular communication of research results. On the contrary, "Portuguese" and Portuguese tend to be conflated as one and the same thing. For those of us who are not familiar with the research procedures of population geneticists, it is easy, then, to get the false impression that our individual genomes contain some kind of mysterious ethnic code that can be unveiled by DNA analyses performed by wizards in white overalls. But in reality, there is no such thing.

In medical studies, the convenient convention of describing populations in terms of present-day national or ethnic identity may be useful because it can help detect statistically increased risks for certain diseases among people with a common ancestry, who in modern society are also likely to share nationality or ascribed ethnicity. But what are the potentials and pitfalls of applying the same concept of population to people in prehistoric times?

The Nation

Most people who practice or take interest in genetics today live in societies where the modern concepts of national or ethnic identity are so deeply rooted that they seem natural and universal. But in a broader historical perspective, the idea of the nation as we know it is a recent and quite marginal phenomenon. The American historian and political scientist Benedict Anderson has famously described the modern nation as an *imagined community*. In the imagined community of a modern nation, people share a sense of belonging and relatedness with millions of fellow citizens whom they have never met and with whom they share no form of personal relations. Anderson connects this concept of the nation with the breakthrough of a set of communication technologies, including railways, calendars, newspapers, and mass-printed novels and popular magazines, in the nineteenth century, precisely as modern democratic nation-states were replacing earlier forms of statehood centered on small ruling classes of royalty and aristocracy. In the service of the nation, the new mass media would ideally reach people of all classes in every corner of a national territory to allow them to share the same information and engage in the same stories and images. Included in the new nation-making technologies were also museums and history textbooks. They united the broad masses of citizens around a common story of origin of the nation, which was typically rooted in prehistoric times and had a successful development through periods of hardship toward a prosperous present and bright future for the nation. By such means, the nation has since the nineteenth century been "conceived as a solid community moving steadily down (or up) history."[25]

Few of us find reason to question the modern concept of the nation in our everyday lives. But when we set out to investigate and tell stories about an ancient past where there were no railways and no mass media, we must try to imagine not only a world that was free from passports and border controls, but also one where the sense of belonging and community was entirely different. In this context, the application of modern population genetics on samples of ancient human DNA will inevitably entail problems. As a field relying on databases of modern DNA, its core methodology tends to impose a population structure similar to that of the modern nation-state on prehistoric people and societies. This is concerning for several reasons.

Not only are modern concepts of national and ethnic identity in all probability not applicable to prehistoric society, but also the unpredictability of aDNA sampling tends to undermine statistical models developed for modern populations, which are based on the principle of random sampling. In aDNA studies you have to work with the samples you happen to get, unlike modern reference databases that have been built on controlled sampling criteria. The availability of aDNA samples depends on a plethora of unknown and uncontrollable factors, such as socially and culturally defined burial practices, local soil conditions for bone preservation, and the slim chances of human remains being picked up in an archaeological excavation and thereafter being stored in a way that preserves organic material, as well as a range of present-day political, judicial, and ethical issues that may allow or forbid DNA analysis. Add to that the methodological restrictions, and it becomes clear that far from all aDNA samples analyzed will result in successful genome-wide sequencing.

Conclusions drawn from studies of 340 individual genomes across two continents and 12,000 years are therefore bound to be frail. Think of it as a jigsaw puzzle where you have 100 pieces at hand, while 100 million pieces are missing. Then imagine that the pieces you have are not necessarily randomly distributed, but could all be gathered in one remote corner of the picture, and you do not even know which one. In other words, even if a single genome can indeed be analyzed to indicate biological relations beyond the individual to whom it belongs, we are left with the fact that we are missing almost all of the actual picture, and we do not know which parts are missing. We simply have to assume (especially if we apply the principle of Occam's razor) that the missing pieces look like what we would expect, and then we tend to depart from modern society and our expectations of prehistoric society as norm.[26]

In aDNA studies, one or a few ancient genomes often come to act as proxy for an entire population, mathematically modeled in the form of a modern population where people tend to meet and mate close to home. But how can we know that this was also the case for people who lived thousands of years ago? Can we trust that the distribution of SNPs in the genomes of ancient people indicate the same kinds of social relations as in a modern nation-state? Well, we cannot be sure, but given the historical specificity of the modern nation-state, it is more than likely that prehistoric people's sense of community and belonging was *not* the same as our own. Instead of staying open to the endless possibilities of conceptualizing kinship relations in ancient societies, an archaeology relying on population genetics thus tends to straitjacket prehistoric people into social relations that follow the norms of modern nation-states.[27]

Ethnonyms

Modeling populations from aDNA is obviously a very different business from the construction of controlled databases of modern genomes. Yet the conclusions drawn are presented with a similar bold imagery and terminology, speaking of bounded populations that move and stay, live in isolation and interbreed with other bounded populations. These populations are often presented as real groups of people with names that allude to national identity, such as "Ancient Beringians" or "Yellow River population" in our two example studies. In reality they are figments of statistical modeling, and their names are convenient conventions with little or no relevance to their ancient contexts. Some populations, like "UPopA," have not been created from actual human remains but are mere virtual constructions that aim to fill gaps in the computer puzzle. Others are represented by a proxy of one existing individual genome, such as that of "Anzick-1," the remains of an infant who lived around 13,000 years ago, which were found accidentally in 1968 during construction work at a ranch owned by the Anzick family in Montana. The name "Ancient Beringians" is derived from Beringia, which was named in 1937 after the Danish-born eighteenth-century cartographer, Vitus Jonassen Bering. It goes without saying that these ethnonyms were of no relevance to the ancient people to whom they are applied. Rather, they are convenient conventions created for our own times. In the cases described above, these labels are of anecdotal interest and arguably have little consequence. However, in cases where the borrowed ethnonyms have social and political relevance today (such as

Han Chinese, Aryans, or Vikings),[28] there may be serious consequences in the form of politically potent narratives in which the meanings and values of identity markers in our own time are projected onto ancient people who likely conceived of their identities in quite different terms. Moreover, it is very unlikely that the "Anzick" or "Ancient Beringians" saw themselves as single, unified peoples over millennia.[29]

Some geneticists have been clear about the caveats of the terminology. Jennifer Raff, for example, admits in her recent book *Origins: A Genetic History of the Americas* that "paleogeneticists often talk about 'a people' based on information from a single genome, while also recognizing that this is a ridiculous characterization."[30] This may be self-evident for specialists in genetic science, but I daresay the crucial difference between a real group of people and "a people" is lost on most nonspecialists. On the contrary—and here I speak from my own experience—we tend to see real people on the move when we see maps with arrows labeled "Anzick" and "Ancient Beringians" spanning continents and millennia (fig. 4). But Raff is absolutely right. Unlike those of earlier culture-historical archaeology, these arrows mostly represent people whose lives left no observable material traces, only shadows in the form of SNPs in other people's genomes. It is essentially a computer game. The movements of virtually constructed populations cannot be empirically observed, only calculated as the single most likely—or least unlikely—explanation for the genetic relations that can be observed between the analyzed ancient genomes, and modern reference genomes. With computer models and algorithms rooted in population genetics, the most likely explanations for their relations have been estimated as population movements in time and space.

The arrows are illustrations of these virtual, computer-generated population movements. They look very similar to those of culture-historical archaeology, but instead of a pots-equal-people paradigm, we could talk about a paradigm of SNPs-equal-people. Unlike the pots-equal-people paradigm, where the manufactured form and fabric of artifacts formed the basis for categorization, the SNPs-equal-people paradigm categorizes people based on minuscule components of their body fabric. With the shift from pots to SNPs, the focus of identity categorization has thus also shifted, from ancient people's *doings* to their material *beings*.

Migration Stories

The crucial difference between a genetic population and an ancient "people," along with the difficulties and caveats involved in working with ancient DNA, call for caution and transparency in the communication of research results. Yet stories of ancient migrations that have recently figured in popular media bear few if any signs of caution regarding the frailty of aDNA sampling and the limitations of statistical analyses of prehistoric materials. On the contrary, they speak confidently of coherent groups of people resolutely moving across continents. In what often appear like intentional colonial conquests, ancient migrants overcome challenges and reach new lands where they replace or interbreed with earlier settlers.[31] With suggestive words and imagery, migrating people are endowed with outstanding collective characteristics: for instance, the Bronze Age Yamnaya culture being described as "a fascinating people," "migrants from Russia," and "the most murderous people of all time."[32]

People of the Yamnaya culture (also known as the Pit Grave, or Ochre Grave, culture), who inhabited the Pontic-Caspian steppe around 5,000 years ago, have been featured as protagonists in some of the most spectacular migration stories that have yet emerged from aDNA research. A series of studies from 2015 onward, including one with the title "Massive Migration from the Steppe Was a Source for Indo-European Languages in Europe,"[33] have generated a wave of popular stories in which tall, strong-bodied Yamnaya men emerged from the Russian steppes and "galloped across Europe, raping and pillaging as they went."[34] Riding the same wave, reconstruction images and television documentaries have portrayed Yamnaya people as stern-looking, muscular men with a dominating appearance (fig. 5).

The truth is that we will never know what the people we now call Yamnaya looked like, and we do not know if they indeed saw themselves as one coherent group of people. Nonetheless, in history-writing based on aDNA, they have been portrayed as a bounded community associated with "'Caucasian' genetic input" and the "spread of Indo-European languages,"[35] which are words and narratives connected with early twentieth-century history-writing focusing on the origin of "Aryans."

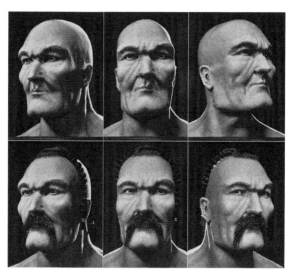

Figure 5. Reconstruction of Yamnaya man by Wojciech Pędzich and Robert Moly-
neaux, CC BY-SA 4.0, via Wikimedia Commons.

The actual genetic facts supporting the stories and images of Yamnaya people are that SNPs first identified in genomes of individuals
associated with the archaeological (pots-equal-people) Yamnaya culture in the area of the Pontic-Caspian steppe at a later date appeared
in the genomes of individuals in central and western Europe, where
they did not occur before. The traces are clearer in comparisons of
Y-chromosome DNA than of mitochondrial DNA, which indicates that
more male than female genetic material has moved westward. Comparisons with ancient human remains of different dates as well as with
modern reference databases indicate that the first major movement
of DNA from the Pontic-Caspian steppe to central Europe occurred
around 4,500 years ago, and that from there it spread quite quickly
and can still be traced in genomes of people with European ancestry.

In no way can the narrative and visual elements of the stories told
about Yamnaya—as a mob of muscular thugs and rapists galloping
across Europe—be motivated by this set of genetic data. Archaeological data can add some pieces; for example, that individuals found in
Yamnaya graves had a larger average body size than hunter-gatherers
in central Europe, and that Yamnaya graves indicate an elite burial

tradition that included grave offerings of horses, wagons, and weaponry. Archaeological findings also suggest cultural changes in central and western Europe around 4,500 years ago, which could be explained by the arrival of people carrying different cultural traditions. Still, most of these spirited stories are pure fabulation.

Many of the fables have arguably been told by enthusiastic popular writers and journalists aspiring to newsworthy features, but they have been enabled by the statements of scholars and scientists. A speculative statement about genocide by archaeologist Kristian Kristiansen, for example, inspired the depiction of Yamnaya as the "most murderous people of all time."[36] And geneticist David Reich, whose lab was responsible for the "massive migrations" study, writes that the few Y-chromosome variations found among Yamnaya genomes "shows that a limited number of males must have been extraordinarily successful in spreading their genes," and that "the preponderance of male ancestry coming from the steppe implies that male descendants of the Yamnaya with political and social power were more successful at competing for local mates."[37] While this is not identical to the narrative speaking of Yamnaya as a band of thugs and rapists galloping across Europe, it sanctions a story of male sexual dominance and passive female compliance—none of which is evident from studies of aDNA.

We must first acknowledge that an evident movement of SNPs (or pots) does not equal the exact same movement of people. People have no doubt moved without carrying pots, and without leaving any SNP traces. And both pots and SNPs can move without their owners. In the case of Bronze Age Europe, the observation that Y-chromosome DNA associated with Yamnaya occurs more frequently in central Europe than mitochondrial DNA can therefore be explained by a number of different scenarios. One would be the equally disturbing story of violent matriarchs in central Europe trafficking male sex slaves from the East for the pleasure of themselves and their sisters. Another explanation that would render the same genetic imprint would be if women had a long-standing tradition of regular journeys to the eastern steppes, where they found men of their liking with whom they mated before returning to their home communities. There are also a whole range of other, less spectacular explanations for the observed genetic patterns. I am not suggesting that any of these stories are true, but they are just as likely if we depart from genetic evidence alone. The point is that the stories about Yamnaya men successfully spreading their genes across Europe by means of violence or man-to-man competition over females—narratives obviously not born in any laboratory test tube—are presented as if they were. The apparent immutability of DNA evidence gives credibility to these stories, no matter how frail the samples or how speculative the explanations.

✕

In his 2011 novel, *The Last Gift*, Abdulrazak Gurnah writes the story of Abbas, a secretive immigrant from Zanzibar to England. Having suffered a sudden stroke in his home in Norwich, the aging Abbas thinks back on his life and reveals unsettling details about his youth in Zanzibar to his wife, Maryam, herself a foundling with unknown birth parents, and their two grown children. Abbas reflects with frustration on his wife's reactions:

> But she says our children are here, in a strange place, and all we have given them are bewildering stories about who we are. She thinks it makes them unsure and afraid about themselves. It makes them lose confidence, she says. As if we should be full of confidence all the time. As if we can know everything we want to know. [38]

The story of Abbas and his family revolves around the situation of first- and second-generation immigrants in search of solid origins to remedy sentiments of nonbelonging and in-betweenness. It is also a story about a society dealing with a history of colonial domination, where social hierarchies hinging on issues of race and migration have become part of the constitution of the nation. As such, Abbas's story epitomizes a current global predicament, here described by the RAND Corporation:

> Every year millions of people leave their homelands to start their lives somewhere else. Some seek to find a promising new job, strengthen family ties, or engage in new cultural opportunities. Some are seeking relief from crushing poverty or a lack of economic prospects. Others are fleeing war or persecution. Still others are escaping the effects of a long-term drought, a devastating hurricane, or some other kind of climate change-related disaster. [39]

In our present world the reasons for deliberate or forced migration are manifold. Most of us have personal experiences or family histories of dramatic migrations. We have all seen news features showing families walking in caravans along endless roads and anxious children clinging to older siblings in vast tent camps. We have heard heartbreaking news of boat catastrophes and read encouraging success stories of brave people having walked through mountain passes and hidden in car trunks to reach a new homeland where they have managed to build a secure and prosperous future for themselves and their loved ones.

In contrast to stories of Yamnaya migrations, which stand out with their idealization of Caucasian male supremacy and sexual violence, most aDNA migration stories have storylines similar to migration stories in our own times. They invoke refugee families in caravans and camps, or brave and fortunate migrants overcoming hardship to find a new homeland where they build a prosperous future and eventually become our own ancestors.

From the evolution of our species we have, for example, stories of Denisovans, scientifically hypothesized as an extinct hominin species[40] after analyses of mitochondrial and nuclear DNA from a few small pieces of bone and teeth found in different locations in Russia and East Asia.[41] There have been no actual finds of more complete Denisova remains, but comparisons of SNPs suggest that the few known pieces of bone belong to archaic human beings whose DNA does not conform to the profiles of *Homo sapiens sapiens* or Neanderthals. The anomaly population was labeled "Denisova" after the place of the first find in 2008, in the Denisova cave in Siberia. Calculations indicate that "Denisova" individuals existed across parts of Asia and Oceania over hundreds of thousands of years, and have left SNP traces in the genomes of modern humans who are now spread across the world, from Iceland to Papua New Guinea. "Denisovans" could thus best be described as a virtual piece in the larger puzzle of human evolution, and with every new find the picture of their relations to modern humans and other hominin species has been further complicated.[42] Hence, as far as I can see, geneticists have not *discovered* a new hominin population or species—what they discovered was a piece of bone with DNA that did not fit their previous models of hominin species. And then they *invented* the "Denisova" population to fill the gap.[43]

This is how the science of population genetics works, so it is not a matter of poor or flawed science. But for most nonspecialists, the distinction between a "population" as a technical term in population genetics, and a population in lay terms (as in the population of a small town, or the current Swedish population) tends to be blurred. Hence we have seen images emerging of Denisovans as a cohesive group of people, described as "an elusive bunch" and a "mysterious crowd,"[44] migrating across continents and mating with other hominin groups they encountered on the way. Such meetings have been depicted in serious popular science magazines like *National Geographic* with suggestive images of classic cave man–style bands, with tangled hair, loincloths, and cudgels, staring at each other across empty landscapes.[45] There are countless examples of similar depictions, in text and images, of Denisovans as a "people," which is something fundamentally different from a "population" as a technical term denoting a statistical probability. As anthropologist Magnus Fiskesjö has pointed out, complex processes of hominin interaction across hundreds of thousands of years have thus been compressed and spun into images and narratives of "groups of people packing their bags to move across the landscape."[46]

Collapsing Time

Recent aDNA studies have inspired plenty of stories of people on the move in prehistoric times, from the "epic migration" of modern humans leaving Africa 60,000 years ago[47] to "large-scale migrations" to the British Isles in the Bronze Age.[48] All these stories essentially rely on time-compressing narrative maneuvers.[49] In studies of early human evolution, the time span of "migrations" can be thousands of years. Studies of later periods like the Bronze Age can speak about movements of DNA in a time frame of a few hundred years, but that is as narrow as it gets. In the recent studies of the peopling of the Americas, population movements are framed within several thousand years, with the exception of the "Southern Native American" population, whose SNP profiles appear to have spread from the northern to the southern part of North America in a few hundred years. Such "leapfrog" movement is exceptional in the perspective of aDNA studies. "These early populations are really blasting across the continent," says archaeologist David Meltzer in an interview, seconded by geneticist Eske Willerslev:

"As soon as they get south of the continental ice caps, they're exploding and occupying the land."[50]

However, if we once again switch perspectives from modern genetic science to actual prehistoric people, we get a different picture. The distance from Beringia to present-day Mexico is around 4,400 miles, and from anthropological research we know that a nomad community easily covers a distance of 10 to 12 miles per day, just living their ordinary life. At that comfortable pace, it would take about a year to get from the northern to the southern end of North America. Then consider the story of fifty-eight-year-old Cargo Harrison, who alone walked the 14,000 miles from Argentina to Alaska—an endeavor in harsh climates that reportedly included a heart attack, a torn tendon, and a fight with a grizzly bear—in 530 days.[51] If we instead say that it took a few hundred years to cover the distance from Beringia to Mexico, which the genetic analyses indicate, we are talking about an average movement of 50 meters (around 54 yards) per day, or 18 kilometers (about 11 miles) per year. For people with a nomadic or semi-nomadic lifestyle, that is hardly blasting speed. Nor would Cargo Harrison be intimidated.

In fact, the evident movement of SNPs from one end of North America to the other within a couple of hundred years could easily have happened without any sense of drama for the prehistoric people involved. They could comfortably have moved to and fro across the continent many, many times within that time frame. But with narrative maneuvers that collapse time, the most sensational stories have emerged to illustrate the process:

> The story this rather dry genetic evidence reveals is breathtaking when you stop to think about it: a small group of people survived one of the deadliest climate episodes in all of human evolutionary history through a combination of luck and ingenuity. They established themselves in a homeland, from which their descendants—hoping to make a new and better life for themselves—ventured out to explore. These descendants found new lands beyond their wildest expectations, entire continents (possibly) devoid of people, lands to which they quickly adapted and developed new ties.[52]

The excerpt is from the recent bestselling book, *Origins: A Genetic History of the Americas*, by geneticist Jennifer Raff. We recognize the storyline here and elsewhere not only from stories of migration in our own times, but also from a previous generation of stories of human evolution. In the book *Narratives of Human Evolution*, bioanthropologist Misia Landau analyzes popular science writings on human evolution from the nineteenth and twentieth centuries, and finds them curiously united around a common narrative, reminiscent of heroic tales in early European literature and folklore. The common story centers on a humble hero (small group of people) departing on an adventurous journey where they face obstacles (a deadly climate episode) and receive help (luck and ingenuity) to overcome them, before they reach their destination (new homeland) and arrive at a higher state (descendants colonizing a new continent).[53]

We see the same narrative structure guiding many migration stories that have sprung from recent aDNA research. The coherent storyline requires a coherent protagonist and a sense of purpose in their movement. In these narratives, therefore, we see populations (which were created as convenient conventions in genetic analyses) acting as unified protagonists, as real groups of people on the move. With a narrative format that ignores real time spans of hundreds or thousands of years, they seem to be blasting, galloping, or just traveling toward their destinations. Hence, scientific papers on prehistoric migration routinely speak of "routes" and "journeys." On eight occasions in our example study of the Americas, population movements across continents over the course of 12,000 years are described in terms of "travel," implying that there was a sense of purpose to prehistoric people's movements, organized around concepts such as origin and destination. This must be regarded as pure narrative invention to make a good story, with no relevance to prehistoric reality.[54] And the arrow works as a visual reinforcement of this sense of purpose in the movement from origin to destination. Gillian Fuller writes, in a semiotic study of arrows at airports:

> [The] arrow doesn't just stabilise the person into "the traveller" with concomitant predicable paths and contractual responsibilities, it also determines specific procedures for movement, for transforming our relationships and personal status. In a world where forward movement is privileged [. . .] the arrow is a trope as well as a tool in this "supermodern" world of constant transit.[55]

Carrying such connotations, arrows on maps indicating ancient people's movements come with a lot of baggage in terms of modern discourse. Of course, it is possible that prehistoric people at some point packed their bags and moved across continents in dramatic journeys to new homelands—we will never know for sure. But the archaeological context does not give any clues, and genetic evidence offers no clear support for such dramatic images and narratives. They seem rather to be inspired by stories of travel and migration in our own times, and the tradition of telling heroic tales in studies of human evolution.

The Politics of Ancient Migrations

What is the problem, you may ask, with a bit of playful speculation and good storytelling? Well, if you maintain, as does British geneticist Mark Thomas, that "science isn't about telling stories, it's about testing them,"[56] then most of this enterprise would not count as science. From the perspective of archaeology, however, storytelling and speculation are not, in themselves, much of an issue. From an ethical standpoint, if we want to respect and defend the integrity of ancient people, we must admit that we tell stories to make sense of the tiny fragments (be they things or molecules) that are left from ancient times. We must also admit that our stories include elements of speculation, because we cannot ask dead people if we got it right. So the problem is not speculation itself, but rather the presentation of speculative stories as if they were built entirely on solid facts. We see in the cases discussed above how the illusory truth effect of DNA allows more or less wild speculations to be presented as full and final descriptions of prehistoric reality. Unlike archaeological stories that are (at best) transparently resting on speculation and imperfect interpretations, the truth effect of DNA offers a seal of closure to speculative stories and thus allows them to have strong impact on official history-writing and public understanding of prehistoric reality.

In circular arguments enforced by statistical models and algorithms, the grand stories of aDNA tend to confirm predicaments of our present modern society as natural and essential to human experience through all times. They speak of prehistoric populations as if they were modern communities, and impose on ancient people sentiments of national belonging and desires for bounded homelands. As such, the turn to aDNA as a basis of grand storytelling robs archaeology of what I see as its main virtue: the potential of the prehistoric past to surprise us, to tickle us as it challenges what we have taken for granted, and to leave us humbled when we realize how little we can actually know.[57] In these stories echo instead the words of the frustrated Abbas: "As if we should be full of confidence all the time. As if we can know everything we want to know."

In our own times of migration, stories of ancient people's movements hold a political potential that brings us back to the arrows. The recent surge of aDNA studies has coincided with political developments that once again speak of nations in terms of pure ethnic or racial communities. Hence, paraphrasing the American evolutionary biologist C. Brandon Ogbunugafor, "the unfounded optimism" inherent in grand-gesture storytelling based on aDNA "might be harmless in a vacuum, but it is pernicious in our universe."[58] In the context of current political developments, it is hardly surprising that the tales of Yamnaya have been picked up as foundational stories of the Aryan race by extreme-right Nazi-inspired groups celebrating ideals of white male supremacy and sexual dominance.[59] Nor should it come as a surprise that aDNA stories with arrows illustrating the origins, historical migrations, and ultimate homelands of prehistoric populations have been taken up for political purposes by nationalist as well as Indigenous leaders in contexts as disparate as China, Greece, Israel, and the United States.[60] In the next chapter we will take a closer look into the concept of ancestry, which connects prehistoric people with communities and political claims in the present.

FURTHER READING

Furholt, Martin. "Massive Migrations? The Impact of Recent aDNA Studies on Our View of Third Millennium Europe." *European Journal of Archaeology* 21, no. 2 (2018): 159–91.

Hakenbeck, Susanne E. "Genetics, Archaeology and the Far Right: An Unholy Trinity." *World Archaeology* 51, no. 4 (2019): 517–27.

Landau, Misia. *Narratives of Human Evolution*. New Haven: Yale University Press, 1993.

Raff, Jennifer. *Origins: A Genetic History of the Americas*. New York: Twelve Books, 2022.

3: A Family Tree of Everyone

For as long as I can remember, I have known that I am a descendant of Saint Bridget (1303–73), the Christian mystic who had eight children, founded the Bridgettine Order, and became one of five patron saints in Europe. I know, because someone has traced my grandfather Bengt's lineage through centuries of archival documents, all the way back to Bridget in the fourteenth century. It was always a good story since I am not exactly a saint myself, so when my colleagues started talking about amazing new technologies to analyze ancient DNA, I immediately saw an opportunity to prove my saintly relations. Surely there would be a piece of life code from Saint Bridget to be found in my DNA?

I never got the chance to try, because recent investigations of Bridget's grave showed that the remains in it were not likely to be hers. And now I know anyway that such an attempt would have been in vain. An individual ancestor as far back as the fourteenth century would be responsible for a few hundred, at most, of the 6 billion base molecules that are contained in my genome. Those few base molecules would be spread out randomly and recombined, moreover, in what is likely to be a degraded sample of aDNA, so they would be practically impossible to locate. The paper trail that leads from my grandfather back to the fourteenth century may have shortcomings, but it is nonetheless a much stronger piece of evidence than a DNA test will ever be. The idea was thus a complete failure, but it has taught me something about the allure of DNA. I know how easy it is to be seduced by the illusory truth effect of DNA as a simple and accessible key to our own ancestry.

Ancestry As Capital

If you send a spit sample to the company iGENEA and pay $509 for a Premium Package, you can find out if you are related to Bernie Sanders.[1] Four out of ten people with family history in Europe will learn that they are, because they share the same haplogroup in their mitochondrial DNA.[2] In genetic lingo, a haplogroup is a result of a small shift of base molecules during the copying process: a mutation, and in this case one that happened in the mitochondrial DNA of a woman who geneticists believe lived in the region of present-day Syria some 25,000 years ago. She herself would not have noticed any effects of it, but since the mitochondrial DNA does not recombine and is passed down intact on the maternal side, all her descendants on the female lines will carry a trace of this mutation, which geneticists have designated as "haplogroup H." If you are one of these people, the iGENEA company can reveal that you are also related to Queen Victoria, Saint Luke the Evangelist, and Warren Buffett.

This might feel like an exciting (or upsetting) revelation, but it doesn't say much of value if you already knew you had ancestors in Europe on your mother's side. Think of it as a thread that runs from you to your mother, to your maternal grandmother, and on to her mother, all the way back to the woman with the mutation 25,000 years ago. If you follow only that thread, you will already have lost sight of three of your grandparents, and when you get back to the eighteenth century, you will be completely blind to more than 1,000 ancestors in each generation—all of whom have contributed just as much to the coding parts of your genome as the one ancestor that you have your eyes on. And likewise, if you think of all the threads that spread out from the woman with the mutation—through all daughters of her daughters, and their daughters in turn, on and on through generations—it will soon be thousands of threads with people who have little in common but a single ancestor thousands of years ago. So, in fact, a test with focus only on haplogroups does not say much about your relation to Senator Sanders, either. If Bernie had, say, a half sister with a mother who did not happen to be on a "haplogroup H" thread, she would not register as his relative in this test—although she in fact has a much closer genetic relation to him than Queen Victoria, who would turn out as a relative. And ultimately, we are all related, so why does it matter, anyway?

In an oft-cited article in the *Guardian*, the British geneticist Mark Thomas has dismissed this kind of enterprise as "genetic astrology," explaining that from a scientific point of view ancestry is something "complicated and very messy."[3] Alongside Thomas's illuminating article, there are masses of available and accessible texts by scientists, scholars, and journalists explaining why a genetic ancestry test in almost every case is a waste of time and money.[4] The exception would be for someone in search of lost family members, or with specific questions regarding known ancestors, who uses DNA in combination with archival documents and family histories to do a targeted search by piecing together a variety of information sources.[5] But this requires considerable investment in time and research skills, so the vast majority of people who have paid $509 and sent a tube of spit off with the postal service will have had information equivalent to a molecular horoscope in return. This begs the question: If the information is out there and the scam is revealed, how can businesses of genetic ancestry testing continue to flourish?[6]

It is actually quite easily explained if we shift our focus away from the scientific quality of the test results and instead consider the illusory truth effect of DNA as evidence, combined with ancestry as a kind of capital with social and political value. Think, for example, about Massachusetts senator Elizabeth Warren's claim of Cherokee ancestry, the subsequent DNA test to prove it, and the bizarre assertion that these claims had no impact on her career.[7] Or Donald Trump, who mocked Warren's claim, while having for years lied about his own "Swedish" family history in Europe (Trump's father, whose parents had immigrated to the United States from Germany, realized the liability of German ancestry when he built his real estate business in the years after World War II, and picked a more convenient story).[8] Or, indeed, media celebrity Oprah Winfrey, who took at least two DNA tests to research her ancestry (one showed links to South African Zulu people, and a second instead showed relations with Kpelle people in Liberia),[9] and said that "it was absolutely empowering to know the journey of my entire family."[10] In Europe, the London *Times* made front-page news of the story that Prince William "will be Britain's first king to have proven Indian ancestry" (based on DNA samples from distant relatives of his mother, Princess Diana, who reportedly had a mitochondrial haplogroup that was most common among people in South Asia),[11] and in South Africa you can book an "Ancestry Experience" as part of your visit to the Origins Centre Museum in Johannesburg and take a DNA test to see if you are related to Nelson Mandela (who was reportedly related to San people by the mitochondrial haplogroup L1).[12]

In all these examples we see ancestry being forged and used as a foundation and anchor of a desired self-image. Mandela's roots in a historically oppressed native community, Warren's failed attempts to be similarly rooted, and Prince William as an icon of an ethnically diverse Great Britain: all these examples point to the value of ancestry as social and political capital, and to the common understanding that DNA has the ultimate power to prove it. However, the strands that point toward these desired images have been picked and teased out of a plethora of alternative and often more obvious ancestral connections. Oprah's case shows how the reliance on different population-genetic databases for comparison (as discussed in chapter 2) tends to give different results with different companies. In the cases of Prince William and Senator Warren, all the fuss and headlines have been about the potential ethnic identity of one distant relative among millions. So, rather than a contribution of high-quality scientific data, here we see DNA used primarily as a narrative device—as a seal of proof on a story that is cherry-picked so that someone can acquire a desired ancestral capital.[13]

The Importance of Context

Celebrity examples aside, ancestry is for most people a serious matter that is intimately entwined in how we define kin and belonging. This is also to a high degree relative to cultural, historical, legal, and bureaucratic contexts. For example, in the context of my native language and the national apparatus I am part of as a Swedish citizen, the word "race" is rarely used in public discourse as a valid category of identity.[14] In Sweden there has been much interest in the use of DNA in genealogical research, but then mostly to fill in specific blanks and find unknown individuals in the family tree. The concept of race is rarely considered important in these pursuits, whereas, for example, in Singapore, South Africa, or the United States, the word and concept of "race" is of great importance because it is at the heart of the nation's history and bureaucratic apparatus. In some contexts, such as China or Laos, the term "ethnicity" is instead paramount, for the nation is conceived as an umbrella over a family of different ethnicities where unruly children ought to be assimilated with the majority ethnic group in power.[15] In yet other contexts, such as the United Kingdom or Canada, national identity has been described in terms of diversity or multiculturalism. There are also in most parts of the world Indigenous peoples, whose identities are defined by a combination of their own cultural traditions and administrative structures, and those of the larger national apparatus in which they are incorporated. For Indigenous peoples in the United States, this includes the broad racial or ethnic definition of "Native American," and a tribal belonging, for example, Hopi or Sioux, that marks the connection to a specific history, tradition, and landscape.

When DNA is used to investigate or establish ancestry, all the entanglements detailed above come into play.[16] In the book *Native American DNA*, Kim TallBear shows how racial and tribal identity is something both complicated and historically contingent. It can be a personal sense of belonging, a spiritual connection, an administrative category deciding rights to health care and homeland, and for some a better opportunity to have their children enrolled at an Ivy League university—all at once.[17] The overlapping concepts of race and tribe both have a relation to DNA in the US context, but TallBear shows how that relation is far from straightforward. More specifically, she discusses how the reference to "blood" or "blood-quantum," as a measurement of tribal identity, has tended to be translated into a language of genetics as what is "in our DNA" or in our "genetic memory."[18] This, she argues, overlooks a much more complex definition of "blood" in relation to the tribe, where it is used as a metaphor of biogenetic relations *and* as a way of life that can be adopted. Rather than indicating a fixation on physical substance, the concept of blood is thus a way to remember and keep record of named ancestors in an intricate genealogical practice.[19]

When studies of aDNA are married to questions of ancestry, the universal scientific language of genetics acquires specific meaning as it is brought into play in all these different contexts. For Indigenous Americans it sticks to, and renegotiates, already existing concepts of blood, race, and tribe. For African Americans, aDNA is most commonly used in investigations of historical sites where objects and human remains are connected to specific African nations or tribes.[20] In the Chinese context, as we will see later in this chapter, work with aDNA contributes to ideas of ethnic identity and becomes relevant to political projects of assimilation. And in the United Kingdom, as we will see in the next chapter, it activates and amplifies fierce debates around the concept of multiculturalism.

In no small measure, quests for ancestral capital and belonging are bringing value to studies of aDNA. Much of the allure and political potential of archaeology lies in the possibility to make connections between past and present. It therefore comes as no surprise that a recent article describes the analysis of ancient human genomes as "a powerful approach for investigating the relationships of people who lived in the past to each other and to people living today."[21] And if we look back on the history of aDNA research, we find it firmly grounded in two previous initiatives to map ancestral relations on a global scale: the Human Genome Diversity Project in the 1990s and the Genographic Project from 2005 to 2020.

The HGDP

The Human Genome Diversity Project, mostly referred to by its acronym HGDP,[22] was initiated in 1991 with a proposal for a new, antiracist, global-scale genetic project in the wake of the Human Genome Project. The new project was fronted by geneticist Luigi Luca Cavalli-Sforza at Stanford University and promised to provide a survey of genetic relations between people, which would reveal "who we are as a species and how we came to be." The way to achieve this was to acquire blood samples from living people in so-called "isolated indigenous populations" and to map their relations by means of DNA analyses, primarily by charting mitochondrial and Y-chromosome haplogroups. The resulting map of relatedness would, according to the proposal, offer a unique window into the evolutionary history of humanity.[23]

Unlike the Human Genome Project, which emphasized the unity of mankind by piecing together DNA fragments from thousands of individuals in one single map of "the human genome," the HGDP foregrounded difference by focusing on the minuscule parts of the genome where human individuals differ. The two projects, though they shared the same official ethos—to provide scientific proof of the nonexistence of human races—and were superficially similar in name, thus had very different objectives. If the Human Genome Project launched genetic *uniformity* in 99.9 percent of the genome as a basis for human unity and an antidote to racism, the HGDP hailed *diversity* in the remaining part as a foundation for the "deep and underlying unity" of humanity.[24]

The 1991 proposal was for a new project, but the HGDP initiative fell back on the initiators' previous research, in particular Cavalli-Sforza's own work in population genetics.[25] Since the 1960s, his research groups had developed statistical models for estimating genetic relatedness between human populations based on analyses of blood collected from "populations" (designated by ethnonyms or national labels such as "Mbuti pygmy," "Swedish Lapp," or "New Guinean")[26] across the world, and had illustrated these virtual relations in the form of dendrograms. Such dendrograms looked much like genealogical diagrams, but they featured genetic populations instead of individuals—ideally forming a family tree for all humankind (fig. 6).[27] The essence of these early works—a global aspiration to encompass all of humanity in a single family tree of genetic relatedness, and to explain current genetic diversity with concepts of historical migration and admixture—was transferred to the new project and formed the backbone of the HGDP.

The HGDP initiators were initially successful at raising funds, and they organized a series of planning workshops in the early 1990s. Before long, however, critical voices were raised. In 1993 a group of anthropologists accused the project of reifying concepts of nineteenth-century racist biology, and soon thereafter members of Indigenous communities launched a series of critical attacks, which culminated in December of the same year when the World Congress of Indigenous Peoples designated the HGDP the "Vampire Project," alluding to the (literal) blood-sucking exploitation of vulnerable people in the name of science.[28]

The criticism was essentially three-pronged. One was about bioethics, where critics pointed to the risks of violation of human rights and exploitation of socially and financially vulnerable people, whose physical substances were used outside of their own control in ways that would only benefit the already rich and powerful. A second questioned the authority to define identity. In a characteristic 1991 *Science* article with the title "A Genetic Survey of Vanishing Peoples," science journalist Leslie Roberts described HGDP as a race against the clock in which "two leaders in genetics and evolution are calling for an urgent effort to collect DNA from rapidly disappearing indigenous populations."

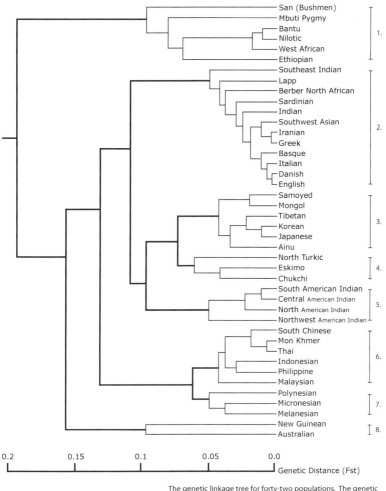

San (Bushmen)
Mbuti Pygmy
Bantu
Nilotic
West African
Ethiopian

1.

Southeast Indian
Lapp
Berber North African
Sardinian
Indian
Southwest Asian
Iranian
Greek
Basque
Italian
Danish
English

2.

Samoyed
Mongol
Tibetan
Korean
Japanese
Ainu

3.

North Turkic
Eskimo
Chukchi

4.

South American Indian
Central American Indian
North American Indian
Northwest American Indian

5.

South Chinese
Mon Khmer
Thai
Indonesian
Philippine
Malaysian

6.

Polynesian
Micronesian
Melanesian

7.

New Guinean
Australian

8.

0.2   0.15   0.1   0.05   0.0
Genetic Distance (Fst)

1. African   5. American
2. Caucasoid   6. Southeast Asian
3. Northeast Asian   7. Pacific Island
4. Arctic   8. Oceanian

The genetic linkage tree for forty-two populations. The genetic distance between any two groups is represented by the total length of the line separating them. Each bar represents one of the eight genetic groups to which all human populations belong. (Adapted from Cavalli-Sforza, Menozzi and Piazza, 1988.)

Figure 6. Dendrogram illustrating relations between populations. Adapted from L. L. Cavalli-Sforza, Paolo Menozzi, and Alberto Piazza, *The History and Geography of Human Genes* (Princeton: Princeton University Press, 1994), and Jonathan Kane, CC BY-SA 4.0, via Wikimedia Commons.

On the first page of the article was a photograph of a bare-breasted woman sitting on the ground with a toddler, along with the figure caption "Vanishing resource."[29] Of course, this kind of representation was upsetting in many ways for the people who were depicted as soon to be extinct. In a statement recalling the much-debated case of Henrietta Lacks, the African American woman whose cancer cells were used without consent or compensation to culture a cell line to the ongoing benefit of global medical research, Rodrigo Contreras, spokesman of the World Council of Indigenous Peoples, said: "The assumption that indigenous people will disappear and their cells will continue helping science for decades is very abhorrent to us."[30] Representatives for Indigenous peoples questioned the authority of the HGDP to assign such identities, accusing the project of "genetic colonialism."[31] The third prong of the criticism pointed to epistemological similarities with colonial science, arguing that the idea of targeting "isolated" and "vanishing" populations because their DNA would offer "a window into the past"[32] was an unsavory revival of racist colonial-era ideals. Anthropologist Jonathan Marks remarked that "the idea that the genes of indigenous peoples are somehow more representatively primitive than those of 'recent, urban' peoples is a holdover from an earlier era, which anthropologists remember with regret."[33]

Such criticism came as a great surprise to the proponents of the HGDP. They were certain their project was neither murky nor exploitative—on the contrary, they maintained that it was a beneficial and powerful weapon against racism[34]—and a frustrated Cavalli-Sforza dismissed the critics as "naïve observers."[35]

Sociologist Jenny Reardon has since argued that the starkly polarized debate cast a veil over the fundamental problem with the HGDP—namely, the premise that the project could generate scientific data that would act as antidote to racist ideologies—as if science was a producer of unbiased truths that could prove ideologies wrong. If anything, says Reardon, the story of the HGDP demonstrates that this is an inadequate view of science. It is in fact a great example of how scientific knowledge, social power, and political ideology are developed in relation to one another.[36]

At the end of the planning period in 1997, the evaluation committee reported that the major funding agencies would no longer endorse the HGDP project. The official motivation was grounded in a questioning of scientific quality,[37] but the volatile politics implicated in the vociferous critique of racism and exploitation undoubtedly had influence on the decision.

The HGDP thus ended after only a few years because it ran out of funding, but it left a vigorous legacy. During its brief existence, it had managed to establish the concept of "admixture" and ancient population splits in the form of dendrograms as viable ways to discuss and make concrete the abstract notion of human diversity. It had moreover established the idea that historic migrations were the primary reason for "admixture" in prehistoric times, and that migrations could therefore explain "who we are as a species and how we came to be."[38]

The Genographic Project

In April 2005, less than a decade after the official burial of the HGDP, the Genographic Project was launched as a private initiative sponsored by the National Geographic Society, the Waitt Foundation, and IBM.[39] The project was conceived and led by the American geneticist and entrepreneur Spencer Wells, who had done his postdoctoral research with Cavalli-Sforza during the years of the HGDP and had made a name for himself as an adventurous and skillful leader of sample-collecting expeditions to remote lands with "isolated populations." The new project, which included Cavalli-Sforza on its international board of experts, was described in almost exactly the same terms as the HGDP: as an "effort to understand the human journey—where we came from and how we got to where we live today."[40] In terms of research methodology as well, the Genographic Project continued on the same path as the HGDP. Haplogroups in the DNA from "indigenous populations" would help to identify "ancestral populations," and historic migrations would be computer-simulated to explain current patterns of genetic diversity. But a clever adjustment was made to the sampling procedure. Instead of targeting people involuntarily sorted by scientists into the category of "isolated population," the Genographic Project used public outreach to encourage people all over the world to engage with the project voluntarily by purchasing a "non-profit" test kit and submitting their DNA samples for analysis. Through their participation, said project advertisements, volunteers would not only gain "deep ancestry insights," but as "citizen scientists" they would also contribute to the greater development of scientific knowledge of human population history.[41] With private funding in place, the Genographic Project had no need for public approval.

The project lasted fifteen years, and at the time of its conclusion in 2020 it had persuaded over 1 million people in 140 countries to provide voluntary DNA samples and contribute to the funding of the project by purchasing test kits. It managed to escape most of the criticism that had sunk the HGDP,[42] although its premises and procedures were much the same. The sharpest edges of the HGDP had been sanded down, and there was no more open talk about isolated populations and vanishing resources. Despite that, there was critique voiced around the Genographic Project that is worth listening to. In this example from a conference panel discussion in 2006,[43] we hear Debra Harry, then representative of the Indigenous Peoples Council on Biocolonialism, and now a professor of Indigenous studies at the University of Nevada:

> The Genographic Project essentially says their genetic analysis will tell us who we are, where we came from, who we are related to, and what the general time-frame is for that kind of information. What they don't tell us is that their research can only be speculative, that the results are inclusive, and they cannot be validated. This is due to the very nature of genetic analysis.[44]

Indeed, the results from a broad-brush effort like the Genographic Project have to be speculative, and questions of who they are, where they came from, and who they are related to cannot be validated for individuals who have contributed funding and samples. Why would scientists advertise research projects with such inflated promises and without revealing the full picture? The answer here, I believe, lies in the quest for samples. An enticing prospect of sensational new knowledge backed up by the all-revealing truth effect of DNA would likely attract more interest and support for the project. Any talk about speculation would very likely have the opposite effect, and fewer people would be interested in submitting samples to the project. And DNA samples are crucial and much-desired assets for population geneticists with grand ambitions.

The key to success for both the HGDP and the Genographic Project was the ability to amass large numbers of DNA samples that could be tested with new computer models and fed into databases for comparison with other samples. In this respect, the Genographic Project was much more successful with its formulation of ancestry as a product for sale and its framing of voluntary contribution as a form of citizen science. Arguably, it also gained purchase on National Geographic's alluring brand of scientific adventure, with Spencer Wells posing as Explorer in Residence on sample-collecting expeditions to faraway lands, as seen, for example, in the award-winning TV documentary *Journey of Man*.[45]

Compared to its officially stated aim to learn "where we came from and how we got to where we live today," the results of the Genographic Project were scanty. As far as I can see, it did not bring any significant new insights into human evolutionary history, and apart from the development of two new genetic research tools (the "GenoChip" and "Geno 2.0 Next Generation"), its main scientific contribution was arguably the accumulated database stored at the DNA Analysis Repository, with its million-plus individual DNA samples.[46]

In terms of public outreach, however, the Genographic Project was an enormous success. It ducked most of the allegations of racism and exploitation that had haunted the HGDP and transferred its ethos into the twenty-first-century consumer market and the sphere of National Geographic, where a wider public got engaged and invested in its cause with voluntary contributions of both DNA samples and funding. So, despite the severe criticism against the HGDP back in the 1990s, the spirit and ambition of Cavalli-Sforza's research—hailing diversity as antidote to racism, creating a family tree of everyone, and explaining current patterns of genetic diversity by means of historic migration—not only survived into the twenty-first century, but got a more favorable reputation and reached a broader audience with the Genographic Project. This, in turn, paved the way for the developments we have seen in aDNA research since 2015:

> Cavalli-Sforza's transformative contribution to the field of genetic studies of human prehistory recalls the story of Moses, a visionary leader [. . .] who created a new template for seeing the world. [. . .] [But] he was not allowed to enter [the promised] land. That privilege had been reserved for his successors.[47]

These lines are from the introduction to the 2018 book *Who We Are and How We Got Here* by David Reich, professor of genetics and frontman of the prolific Reich lab at Harvard University. Reich has been one of the loudest proponents of the "aDNA revolution," and his unreserved celebration of Cavalli-Sforza as a legendary leader of biblical proportions (and self-nomination as successor on the same mission) indicates the importance of Cavalli-Sforza's spirit and ideas on the development of aDNA studies. Several key players in the field, Reich included, are students, or students of students, of Cavalli-Sforza. Many use databases from the HGDP and the Genographic Project for reference and circle around the same concepts—ancestral populations, haplogroups, migration, and admixture—to describe and explain patterns of genetic relatedness between people in prehistoric and present times.[48]

Much like the discussions around the HGDP, most critical debates around aDNA research have concerned bioethics and the effects on Indigenous communities. Researchers in North America and Australia have been most active in these debates, and several projects have been introduced to involve Indigenous people in ancient DNA research.[49] These are important initiatives, but there are a number of apparent legacies from the HGDP that have not attracted as much critical attention when they have resurfaced in current aDNA research.

HGDP Legacies

One such legacy is the portrayal of genetic science as producer of untainted truth that can help overcome ideological and historical biases in archaeology. As Jenny Reardon noted already in relation to the HGDP, this is obviously a false notion. Like all other science, genetics is and has always been developed in tandem with social, political, and economic developments in society at large.[50] In aDNA research, the flawed image of genetic science as an untainted and superior form of knowledge has arguably stood in the way of more nuanced discussions about the benefits and drawbacks of genetic methods in prehistoric research.[51]

A second legacy is the fierce pursuit of DNA samples. Just as in the two previous projects, aDNA researchers who aspire to tell grand stories about the origins of nations, or indeed humanity at large, are in need of large numbers of DNA samples to build databases. Ancient human remains that contain enough genetic material to extract and successfully sequence are rare and hard to come by, and special sampling permits are often required from museums, national repositories, or descendant communities. Ancient DNA thus poses a new "vanishing resource" challenge, which has inspired researchers in the field to talk about ancient human remains as "gold nuggets," and the frenzied hunt for access has been described as a "bone rush" or a "bone bonanza."[52] Both the HGDP and the Genographic Project can teach us something about the value of DNA samples in this context. In both projects the acquisition of samples was of crucial importance, and much of the framing and advertising of the projects' aims were formulated to the end of acquiring large numbers of samples. The same premises apply to aDNA projects with grand ambitions. Researchers and laboratories who are in control of large databases—even if published open access—gain power and prestige, which enables them to maintain or even expand their enterprises. It is reminiscent of nineteenth- and early twentieth-century acquisitions of archaeological artifacts and human remains that were taken in colonial contexts and brought to European museums, where they were used as a kind of professional exchange currency that gave individual archaeologists and collectors professional prestige, and facilitated their continuous enterprises. Such practices have been rightly criticized in postcolonial times, and have in recent decades prompted official apologies in the form of repatriation and restitution. The similar professional value of genetic material is an aspect rarely exposed in open discussions of aDNA research, but one that is bound to have crucial impact on developments of the field. The lessons learned from the HGDP tell us there is trouble ahead if the obvious risks of exploitation are not carefully and openly addressed.[53]

Another predicament that aDNA studies inherited from the HGDP is the reference to "population admixture" as a measure of diversity.[54] As noted by Kim TallBear, in the concept of admixture lurks a notion of genetic purity. "Of course," she writes, "mixing is predicated on purity."[55] It would not be possible to conceptualize a mix of any kind, or talk in meaningful terms about mixture, if there were no unmixed units that first went into it. Hence, the talk about genetic admixture presupposes the existence of genetic purity—even if we do not say it explicitly—and projects that hail diversity as evidence of the nonexistence of pure races at the same time as they define diversity as genetic admixture are bound to be caught in an inescapable paradox.[56]

The most visible legacy of the HGDP, however, is the ongoing scientific project to create a "family tree of everyone."[57] If the arrow is the paramount sign of ancient migrations, the tree is the principal illustration of human ancestry. The dendrogram entered public discourse long before the HGDP, in early nineteenth-century popular renderings of evolution studies.[58] But with the HGDP and the Genographic Project, the traditional genealogical family tree came to merge conceptually with the evolutionary dendrogram, in a tentative "family tree" for all of humanity.[59] Especially with the Genographic Project, which contributed to a surge in the private market for genetic ancestry testing, the dendrogram came to serve as a widespread illustration of the relation between individual and collective. Unlike traditional genealogical family trees, where the collective was the immediate family, the collective was now all of humanity, albeit divided into clear-cut "branches" of ethnic groups and national communities.

The Power of the Tree

The tree illustration, where all of humanity is contained in one single image of relatedness, is impactful with its ostensible message that we all belong together—as one more or less happy family. At the same time, however, the branching tree nurtures notions of essential difference. As is evident from the widespread interest in genetic ancestry testing, people are not content with knowing that they belong to the family tree of humanity. Many are prepared to pay substantial amounts of money and submit their genetic material for analysis with the hope of finding out *exactly where*, on which of the tree's branches, they or their ancestors are located.

We have already discussed haplogroups as one way of mapping relations in a genetic ancestry test. Another common but equally questionable way is to discern "ethnicity." In this context, "ethnicity" is synonymous with genetic "population," which, as we know from chapter 2, is not the same as socially acquired and forged ethnicity, but is rather a statistical estimation of genetic relatedness among samples in a database, and a convenient convention to facilitate communication between geneticists. The "ethnicity" pie chart that is a common result of a genetic ancestry test is thus nothing more than a mathematical calculation of genetic proximity to other customers in the same company's database. But this is not how it is communicated to consumers, who are instead led to believe that there is some kind of ethnic code that has been transferred intact from their ancestors and can be revealed in their DNA. Here, the family tree of everyone sends another clear message. With its discrete and separate branches, it conveys an image of humanity as being irreversibly divided into essential ethnic groups, where ethnicity is predicated on biological ancestral relations.[60]

The family tree of everyone also has a temporal dimension, since it gives the impression of containing all human beings who have ever lived and contributed to the current human gene pool. As such, the tree shows that from a single point of origin, humanity has developed into a multitude of discrete population branches, each pointing toward a future in solitude (or further separation) and ending in its own telos. To an outsider it may seem as if the tree has sprung directly out of the genetic data, but the form is actually generated by the computer programs that have been created and chosen for the analysis. As pointed out by cultural studies scholars Marianne Sommer and Ruth Amstutz, these programs (for example TreeMix and qpGraph) are based on certain presumptions (such as a list of included "populations" and "admixture events") and will always produce results in the given tree form. If you change the parameters in the program and enter the same data, a different figure will appear and create a different pattern of relatedness between the analyzed samples.[61]

This is not to say that population-genetic analyses are bogus, but it is important to know that the statistical computer programs that generate the results we read in scientific papers are not a mechanical and unbiased tool. They are creative and will generate results in certain forms, depending on how the programs are constructed and how the data are formatted. Hence, the family tree should not be regarded as the only possible illustration of human genetic relations. Rather, it has been actively developed and has become standard in aDNA studies,[62] largely because of the legacy of Cavalli-Sforza's work. With its temporal dimension, the family tree acts as a visual confirmation of meaningful relationships between people in the past and present. With its predicated flow from a common ancestral stem to a multitude of discrete branches, it emphasizes the original historical relatedness of humanity at the same time as it nurtures an image of now-living people as essentially divided into separate and genetically identifiable bioethnic groups.[63]

The declared message of Cavalli-Sforza and his HGDP colleagues was that the present is characterized by bioethnic diversity, which will naturally fall back on our common ancestry as a fact with unifying social and political potential. But what happens when this template is transferred to studies of aDNA?

Ancient DNA and the Value of Ancestry

Beyond doubt, the concept of ancestry gives value to aDNA studies. The family-tree figure inherited from Cavalli-Sforza and the HGDP, which leads a continued existence in aDNA studies by virtue of disciplinary convenance and use of the same databases and similar computer programs, thus prefigures the sociopolitical potential of aDNA. It appears to draw absolute and evidential connections between seemingly existing bioethnic groups in the past and the present—connections that can bring both benefits and problems for individuals, communities, and interest groups today.

In many parts of the world, Indigenous peoples have embraced the potential benefits of aDNA and have begun to collaborate with geneticists to get support for repatriation cases. These cases often concern human remains that were taken by colonial powers to build ethnographic or anatomical collections in metropolitan museums and that are now being claimed for restitution and reburial by Indigenous communities. Although the relation between genetic ancestry and tribal identity can be far from straightforward,[64] DNA analyses that can certify ancestral relations between the human remains and the people who now claim them can be of great value for such cases.[65] A great example is the story of Ernie LaPointe and Tatanka Iyotake, also known as Sitting Bull (fig. 7).

LaPointe is a Lakota man who lives in Rapid City, South Dakota. He is the great-grandson of the Hunkpapa Lakota Sioux chief Sitting Bull (1831–90), legendary advocate for Native American ways of life and iconic leader of Indigenous resistance against the US westward expansion and regional gold rush.

The relation to his great-grandfather has always been known to LaPointe, who has had it established by oral history, official documents, and ceremonial confirmation. Empowered by the genealogical as well as spiritual connection, he has published a book, produced a film, and for some years has acted as the authoritative voice of his ancestor.[66] His family has also been involved in struggles over the rights to Sitting

Figure 7. Ernie LaPointe in 2014. Photograph *Eternal Field* by Shane Balkowitsch, CC BY-SA 4.0, via Wikimedia Commons.

Bull's physical remains. At the age of five, LaPointe witnessed the reburial of Sitting Bull's bones after his mother and aunts had authorized an exhumation of the original grave. Half a century later, he became involved in a repatriation case when a braid of hair and a pair of leggings (presumably taken by the medical examiner after Sitting Bull's death) were found in the collections of the Smithsonian Institution in Washington, DC. According to the Native American Graves Protection and Repatriation Act (NAGPRA), claims from lineal descendants had priority over tribal claims, and since official documents suggested that LaPointe and his sisters were the closest living descendants, the braid and leggings were returned to the family in 2007.

There have, however, been many claims on Sitting Bull's legacy, and the exact location of his grave remains an issue of contestation.[67] Ernie LaPointe therefore responded with interest when a Danish team of geneticists approached him shortly after the 2007 repatriation and proposed a DNA analysis of a sample from the hair braid. In an interview, LaPointe explained that although he had no doubt that Sitting Bull was his great-grandfather, a confirmation by means of DNA would give him full control over his legacy: "That will solidify our connection. [. . .] No one can say we are unrelated."[68]

To make sure he had his forefather's consent, LaPointe invited the lead investigator, Eske Willerslev, to South Dakota for a spiritual ceremony led by a medicine man. During the ceremony, Willerslev reported having seen "a blue-green light that ran across his body and into his mouth." This was read as a sign of approval from Sitting Bull's spirit, and the geneticist was allowed to cut a section from the repatriated hair braid and bring it back to his laboratory in Copenhagen. In addition, LaPointe helped Willerslev to acquire more DNA samples for comparison from his family and other living Lakota people.[69]

The analysis went on for fourteen years. When the results were eventually published in 2021, the press release and subsequent popular reporting spoke of a successful operation, and Willerslev said his team was "delighted to find that [the samples] matched."[70] Judging from the scientific article, however, the results were more modest. It had in fact not been possible to extract much genetic material at all from the braid, so in order to get a result the researchers had to invent a whole new "probabilistic method." New computer programs were custom-made to simulate how many of the few identifiable SNPs in the fragmented DNA from the hair would have been likely to have been passed on to Ernie LaPointe if he was indeed Sitting Bull's great-grandson. According to the article, it would in principle be possible to get the same results from the simulation exercise "even if Ernie LaPointe and Sitting Bull were not related," but "this is unlikely." Relying almost entirely on computer statistics, the results were primarily based on assumptions and estimations of the real relations. The article nonetheless concluded that "the results from the real data combined with the simulation results show strong support for Ernie LaPointe being Sitting Bull's great-grandson."[71]

Long before the genetic analysis was even proposed, LaPointe had a strong story and solid evidence of his family relations to Sitting Bull. It is easy to understand the temptation to turn to DNA for further confirmation, and at first glance it seems that is what he got. But if we scratch the surface, we see that the genetic analysis is here little more than a mathematical calculation based on the facts provided by the already existing family history and paper documentation, which in itself is far more robust in terms of evidence than the analysis of severely fragmented DNA. As such, this is a good example of how easily DNA can pass as a stamp of proof on a good story and a desired ancestral relation, even when the quality of the scientific analysis is wanting.[72]

It can indeed be argued that no harm was done in this particular case. Few would question LaPointe's relation to Sitting Bull, and the DNA study only poured icing over an already well-baked cake. But the story calls for vigilance. If the power of DNA as a narrative stamp of proof is equally strong—or in some cases considerably stronger—than its scientific contribution, are there any limits to the stories it can justify? And what kind of storylines will it encourage and validate?

Indigenous Peoples and Genetic Science

There are instances in which prospects of aDNA research have been greeted with skepticism. For example, in New Caledonia, a French overseas collectivity in the Pacific Ocean, national authorities issued in 2017 a three-year moratorium on DNA sampling for historical research because they saw risks that a project aiming to "define the characteristics of the Kanak human genome" and trace their ancient origins would have negative effects on the social order of the Indigenous Kanak, "which has been forged over the centuries by historical processes and contacts between welcoming and welcomed groups of people."[73] In light of the critique against the HGDP and the Genographic Project, this seems like an insightful approach, but it is rare. In most parts of the world, Indigenous communities have rather embraced the possibilities to collaborate with aDNA researchers.[74]

Do Indigenous peoples and tribes need aDNA studies? I am in no position to give a conclusive answer, but my impression is that the answer will vary depending on whom you ask. Some have already turned to aDNA for support in legal cases on repatriation and land claims,[75] and some have found genetics useful in their personal searches for genetic confirmation of ancestral relations to claim tribal belonging. Others have been troubled by the emphasis on biological relatedness if we accept genetics as the one and only measure of true ancestral relations, concerned that such a view will denigrate the much more complex and sophisticated definitions of ancestry and kinship that define the cultures and social identities of many Indigenous peoples.[76] So it should be fair to say that the jury is still out on the question of whether or not Indigenous peoples and tribes need aDNA studies.

Do archaeogeneticists need Indigenous people? Without a shadow of doubt. Following the legacy of the HGDP, in which Indigenous peoples were seen as "isolates of historic interest,"[77] archaeogeneticists with grand ambitions need the consent of Indigenous people to get access to DNA samples, which they need to build databases and their "family tree of everyone." On a smaller scale, some archaeogeneticists have also set out to work with projects relating to the history of Indigenous peoples.[78] In these projects, the earlier mistrust generated by the HGDP among Indigenous communities has posed a major problem for later generations of geneticists working with aDNA. Some have set out to overcome the problems and build trust by creating platforms for education and collaboration.[79] Others have offered narratives as a form of payback for samples—stories of origin where Indigenous people are featured center stage.[80] While this is no doubt a great improvement over previous narratives, where Indigenous people existed only in the margins, if at all, I am not convinced it is a long-term solution to the larger issue of the marginalization of Indigenous people in historic and contemporary society.[81] In fact, it may well generate a double-bind situation, where the cost of collaboration is that key concepts of Indigenous culture and identity are denigrated as artifacts of the past—in contrast to the all-revealing truth machine of genetic science. The only clear winner of such collaboration is the geneticist, who not only gets access to the samples they need, but also gains prestige by posing as benefactor, offering an enticing story of origin (and thus a politically useful stamp of originality) to Indigenous peoples. A far more sensible way forward, especially if we hail diversity as an ideal form of human existence, would be to include a much broader set of voices, listen, and learn from more and different ways to define ancestry.[82] But as long as the academic community makes the choice to support and promote the story of genetic science as a superior form of knowledge when it comes to kinship and cultural history, there seems to be no way out of this double bind.

Considering the fierce competition for samples in the current field of aDNA studies, it seems like a sinister twist of fate that the remains of Sitting Bull, who once fought for the rights of his people against a United States blinded by the prospect of acquiring land and finding gold, and who allegedly predicted that a battle would one day rage over his bones,[83] are today being flown across the world in the context of a competitive bone rush.

An "Unmixed" Individual

In the study of Ernie LaPointe and Sitting Bull, we find another troublesome legacy from the HGDP. Although the research question concerns the individual relationship between LaPointe and Sitting Bull, the scientific paper circles around the concept of population. And among the sampled Lakota, it says, there are five "unmixed individuals." LaPointe is noted as one of these individuals, of whom none is "admixed or interbred."[84] This sounds strange in light of the oft-cited truism among geneticists that there are no human races and that we are all results of "mixing." So shall we understand the description "unmixed individual" and the lack of "interbreeding" in terms of racial purity? Not quite. These terms fall back on the population-genetic concept of "admixture," meaning that the individual has genetic ancestry from only one of the populations applied in this particular study. Like the concept of "population," it is a technical shorthand and a convenient convention to facilitate communication between geneticists. But in a socially sensitive or politically invested situation, this is dangerous language.[85] It also demonstrates the ambiguous attitude toward concepts of race and racial purity in the field of aDNA studies. Just as proponents of the HGDP used models and illustrations of discrete populations to argue that humanity is essentially diverse, the field of aDNA studies is caught in the same paradox when it continues to denote individual human identity in degrees of admixture.

The talk about "unmixed" individuals and bounded "ancestral populations" should not, however, be attributed entirely to the legacy of the HGDP. It has likely lingered to today because these are attractive and powerful concepts when it comes to proving links between ancient people and contemporary communities. As such, they are useful in repatriation cases and other legal processes that speak the language of the law. But when geneticists lend their services to some interest groups—no matter how legitimate or reasonable their claims—by offering results that speak of "unmixed individuals" or scientifically verified links to "ancestral populations," they should be aware that they also fuel the flames of other political movements that take interest in essential racial or ethnic identity.

The Political Potential of the PCA

In the People's Republic of China, there has been an intense interest in the developments of aDNA research. In line with the political agenda of the current regime, which is grounded in the majority Han ethnic group and has a long tradition of registering ethnic minorities as primitive and in need of assimilation to further the prosperous development of the nation,[86] Chinese aDNA research has had a clear ethnic focus. We see it, for example, in a recent article by Vikas Kumar and colleagues[87] denoting the identity of prehistoric populations in the northwestern Xinjiang province, now home to the Uyghur people. Based on a study with genome-wide data from 201 ancient individuals from 39 archaeological sites, the authors conclude that the Bronze and Iron Ages of Xinjiang were characterized by a complex history with "persistent interactions between various populations and cultures." Commentators on *Radio Free Asia* have warned that these results can be used to support the Chinese regime's forced assimilation of Uyghur people—which Amnesty International has reported as "systematic state-organized mass imprisonment, torture and persecution amounting to crimes against humanity" and the United Nations has described as "serious human rights violations"[88]—for which a core argument is that the Uyghur lack their own culture, language, and history in Xinjiang.[89]

It may well be that the archaeologists and geneticists involved in such studies are ignorant of their sociopolitical potentials. Nevertheless, they produce data formatted in figures that are useful for political actors who draw on notions of ethnic essence and continuity between the past and the present. The Principal Component Analysis (PCA) is particularly useful here. A PCA is a statistical calculation of degrees of similarity between chosen sections of sampled genomes. It is presented as a visual diagram with dots representing individual genomes. The dots are often clustered and marked in different shapes or contrasting colors to emphasize the formation of "populations," which are statistically calculated clusters of genetic similarity. Dots that fall out of the common pattern are described as "outliers" and can be dismissed from the analysis. In studies of ancient DNA, it is commonplace for dots from a modern reference database to form an underlying layer with clusters marked by modern ethnic terms (such as the Han, or the Uyghur), with aDNA samples overlapping to show their degree of likeness to the modern clusters (fig. 8).

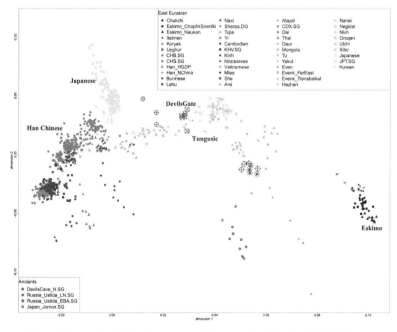

Figure 8. Example of PCA diagram "projecting ancient Devil's Gate, Ustlda and Jomon samples onto the variation of present-day East Eurasians." From Chuan-Chao Wang and Martine Robbeets, "The Homeland of Proto-Tungusic Inferred from Contemporary Words and Ancient Genomes," *Evolutionary Human Sciences* 2, e8 (2020). CC BY-SA 4.0, via Wikimedia Commons.

CHAPTER THREE

The technovisual impact of the PCA should not be underestimated.[90] It recalls a geographical map, where populations are bounded and exclusive, and an overlap between ancient and modern genomes appears to prove clear-cut ancestral relations and anchorage in the same territory. This is not what the PCA actually shows—there are many possible explanations for patterns of overlap, and the patterns themselves rely on the formats of the calculation programs and the parameters of the data put into them—but it is the impression it gives, especially for amateur readers like myself. In the PCA of the Xinjiang study,[91] the Uyghur has been marked as a cluster in the underlying layer, and it appears as if none of the ancient DNA samples overlaps with the modern Uyghur. The clear message is that the Uyghur people have no ancestral links to the Bronze and Iron Age populations in Xinjiang. In contrast, we have the study of ancient migrations by Chao Ning and colleagues, discussed in chapter 2. Their research questions concerned prehistoric developments in and around the Yellow River, which is commonly regarded as the prosperous center of the Han people and the historical breeding ground of the People's Republic of China. In that study, the PCA indicated close genetic relations between ancient samples from the Yellow River region and samples from modern Han Chinese.[92] The Uyghur people were (literally) not even in the picture.

Critical and concerned voices have been raised over the developments in aDNA research in China.[93] Scholars with insight into Chinese media and society have pointed out that state-controlled research institutes and researchers who are featured in national media are bound to present results that align with government policy. It has also been pointed out that non-Chinese scientists engaging in such research projects may be naive or ignorant of the political consequences of the research they endorse.[94]

This calls for reflection, even if we are not working in China. It is quite clear that we cannot regard genetic science as producer of untainted truth. In China and everywhere, it is aligned with social, political, and economic developments in society at large.[95] Tree illustrations and PCA diagrams can thus be put to the service of various political projects, but we are more prone to regard it as politicization if we do not agree with the policies.

Three Takeaways

There are three important takeaways from the discussions in this chapter. One is that when we talk about ancestry in relation to aDNA, we are talking about meaningful relations in the present, not in the past. There were of course sexual interactions between people in ancient times that influenced the genetic makeup of people living today. But the way we make sense of those interactions—in terms of distinct ethnic populations and "admixture events," and in images of family trees and PCA diagrams—is entirely owing to knowledge structures in the present. Moreover, the lesson we can learn from listening to people who have broader and more complex kinship relations—which may, for example, include nonbiological human relations and relations to animals, inanimate objects, places, and historical events[96]—is that ancestry, according to the definitions of genetic science, is extremely restricted, not to say poor. The concept of genetically determined ancestry that is imposed on ancient people as if it were universal is in fact a modern construct much like the nation-state. In all likelihood, it is an inadequate proxy for kinship relations in ancient times.[97]

Second, the words and illustrations that are used to denote and discuss ancestry in aDNA research do not spring neutrally out of pure data. They fall back heavily on the history of the research field, and most importantly on Cavalli-Sforza's projects. The ambitions as well as the predicaments of this earlier research are inscribed in computer programs and visuals like the "family tree of everyone" and PCA diagrams with ethnic clusters. From there, they continue to predicate our understanding of ancestral relations and the evolution of humankind in the form of discrete ethnic groups. Such an outlook facilitates stories that serve the present in different ways, and we may like some of them more than others. We should remember, however, that when we open the door for one, we invite the other as well.

Finally, we have seen several examples of DNA being used as a stamp of proof on a good story of desired ancestral relations. In these examples, the underlying science has been flexible enough to allow for cherry-picking and quite extreme interpretations, which are nevertheless presented as if they rested on rock-hard evidence. If we want to understand the potential uses of aDNA to indicate ancestry, we therefore have to look beyond the actual scientific procedure—because, evidently, that is not what matters the most in these cases—and start taking seriously the use of DNA as a narrative device.

FURTHER READING

Kowal, Emma. *Haunting Biology: Science and Indigeneity in Australia*. Durham, NC: Duke University Press, 2023.

Nash, Catherine. *Genetic Geographies: The Trouble with Ancestry*. Minneapolis: University of Minnesota Press, 2015.

Nelson, Alondra. *The Social Life of DNA: Race, Reparations, and Reconciliation after the Genome*. Boston: Beacon Press, 2016.

Reardon, Jenny. *Race to the Finish: Identity and Governance in an Age of Genomics*. Princeton: Princeton University Press, 2004.

Sommer, Marianne. *History Within: The Science, Culture, and Politics of Bones, Organisms, and Molecules*. Chicago: University of Chicago Press, 2016.

Strand, Daniel, and Anna Källén. "I Am a Viking! DNA, Popular Culture and the Construction of Geneticized Identity." *New Genetics and Society* 40, no. 4 (2021): 520–40.

TallBear, Kim. *Native American DNA: Tribal Belonging and the False Promise of Genetic Science*. Minneapolis: University of Minnesota Press, 2013.

4: Paleopersonalities

"That's incredible," says the young man excitedly. "That's brilliant!" The camera is set on two white men sitting on the stairs of the impressive twelve-columned portico of University College London. One is a British archaeologist, the other a German computer scientist, and together they look down at the screen of a MacBook laptop. The cause of their excitement is an email from a consultant phenotype expert in the United States who has estimated, with a probability of 76.6 percent, that the skin color of Cheddar Man was "dark-to-black."[1]

A Celebrity Caveman

"Cheddar Man" is the modern alias for an individual whose skeleton can be seen on display in the Natural History Museum in London. Judging from his bones, he was a male 166 cm (about 5 foot 5) tall, with good teeth, who was in his early twenties when he died around 10,000 years ago. The time when he lived is called the Mesolithic—the middle Stone Age—and in northern Europe this was the period after the last Ice Age. It was before the use of agriculture in this part of the world, so people ought to have lived from hunting and gathering. But Cheddar Man is not just any Mesolithic bloke. Ever since the first piece of his skull was found in a cave near the village of Cheddar in Somerset, England, in 1903, he has been something of a celebrity.

In the late nineteenth and early twentieth centuries there was a growing public fascination for caves. Caving was established as a sport among the educated and upper classes in Britain, and landowners with entrepreneurial talents saw opportunities to develop cave systems into businesses by opening them for visitors. Among the more ambitious were the Gough family, landowners in the picturesque Mendip Hills near Cheddar. They opened a first cave for paying visitors in 1877, and at the entrance they arranged a kind of museum display of ancient artifacts and bones from extinct animals along with stalactites and stalagmites—all lit up, first by gas and later by electricity. Over the next few decades, they expanded the business with another cave (now known as Gough's Cave), and it was in the process of digging a wider entrance to the new cave that the two brothers Arthur and William Gough found the first pieces of a human skull in December 1903.[2] The brothers were entrepreneurs, not scientists, but they carefully collected all the pieces they could find of the skull, as well as an arm and a leg bone, some ribs, and parts of the pelvis, before they called for professional assistance. An archaeologist and a geologist came and documented the positions of the removed bones, and the rest of the nearly complete skeleton that was left in situ for future investigations. Geologist Henry N. Davies photographed the reassembled fragmentary skull, noting that the face was "much mutilated, and filled with a concrete of cave-earth and calcareous cement" (fig. 9).[3] The skull was then put on display in the cave museum, along with the rest of the collected skeletal remains. In a most remarkable photograph, which was also turned into a postcard, the Gough brothers pose at the opening of the cave sporting impressive mustaches, hats, and waistcoats, with spades leaning on their shoulders. Between them is the reassembled skull of Cheddar Man, piled up on a heap of bones with "a medley of chipped flints" in front.[4]

Now, the late nineteenth century was not only a time for caving. It was also a time when nation-states were competing for power and prestige, and prehistoric archaeology expanded its searches for the earliest traces of humans in Europe. Advances were made, not least in France, where find-places gave names to the earliest archaeological periods (such as the Acheulean and the Magdalenian), and a find of early human remains in the Cro-Magnon rock shelter in 1868 gave name to "the Cro-Magnon Man"—modern man himself. In the imperial rival Great Britain there was quite a bit of frustrated yearning to find equally old and important finds on their own turf.[5] The find in Gough's Cave thus appeared as manna from heaven. The age of the remains was first estimated at a staggering 40,000–80,000 years, and the individual was described not only as "the earliest Englishman" but also "the

CHAPTER FOUR

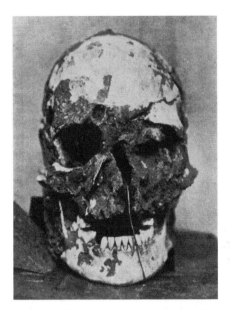

Figure 9. Cheddar Man's skull, reassembled and documented by Henry N. Davies in 1903. From C. G. Seligman and F. G. Parsons, "The Cheddar Man: A Skeleton of Late Palæolithic Date," *Journal of the Royal Anthropological Institute of Great Britain and Ireland* 44 (July–December 1914): 241–63.

oldest human skeleton ever discovered."[6] Such claims would later be confirmed to be greatly exaggerated, but by then Cheddar Man had already become a local and national celebrity, and a golden goose for the Gough family. They continued to expand their cave tourism business, culminating with the 1934 opening of the Caveman Restaurant, a magnificent art deco landmark complete with an Explorer's Café-Bar beside a fishpond on a glass roof. The years around the opening of the new restaurant saw a series of archaeological excavations reported in the press, in which the remainder of the Cheddar Man skeleton was unearthed, along with stone artifacts and animal bones from different parts of the caverns.

Mitochondrial DNA and the Establishment of Relations

The remains of Cheddar Man were eventually moved to London and the Natural History Museum, where they can now be seen on display.

Thanks to the calciferous environment of the cave where they had lain buried for millennia, the remains are remarkably well preserved. Radiocarbon analysis has now estimated that the "Cheddar Man" individual lived around 10,000 years ago, which means that he was far from the first human being to have lived in the current territory of the United Kingdom (where different hominid species could have existed as early as 1 million years ago, and the modern human species for around 40,000 years). Even in Gough's Cave there have been finds of human remains dating several thousand years before those of Cheddar Man. But his celebrity status has lingered, and Cheddar Man is continually referred to as "the oldest almost complete skeleton" found in Britain, or even "the first Brit."[7]

In the late 1990s, the Oxford University professor and popular science personality Bryan Sykes performed a mitochondrial DNA (mtDNA) analysis on the skeleton.[8] This was in the first wave of aDNA studies when focus was on mitochondrial and Y-chromosome DNA, and Sykes describes the research process (which was done as part of a television program) in his popular book, *The Seven Daughters of Eve*. I have not been able to find any published scientific paper on this study, but according to Sykes's own popular account, he eventually succeeded in extracting enough genetic material from one of Cheddar Man's molar teeth to be able to perform an analysis of mtDNA.[9] His interest in Cheddar Man was quite specific. It had little, if anything, to do with Cheddar Man as an individual, or the Mesolithic times when he lived. Sykes was interested in telling—and selling—stories about identity and belonging for people living today, which he did with much success in a number of bestselling books, television programs, and not least in Oxford Ancestry, his own commercial company for genetic ancestry testing.[10] Hence the paramount question for Sykes in the context of this television program/genetic study was whether any haplogroups (specific combinations of polymorphisms that are results of earlier mutations) in Cheddar Man's mtDNA could be matched with people living today. He departed from the seven "daughters of Eve" matrilines he had launched in his popular works, and set out to test pupils in a local Somerset school to find matches. The search was featured in the television program as a kind of scientific detective story, and it was announced a great success when they found a "match" with the school's history teacher, a man named Adrian Targett.

But what does it mean to have a "match" in this context? Well, it basically means that Targett and Cheddar Man both have mtDNA that can be ascribed genetically to a particular haplogroup—U5. This, in turn, shows that they have a common maternal ancestor, whom some geneticists have calculated lived around 55,000 years ago in the region of present-day Russia.[11] Importantly, however, it does not say anything about close relations, since the U5 haplogroup is quite common and shared by as much as 10 percent of all Europeans today. Moreover, an analysis of mtDNA, which is inherited on the maternal side, could not, for obvious practical reasons, be used as an indication that Targett was a direct descendant of Cheddar Man (it does not rule out the possibility but does not support it either). So even if we trust that the sampling and analysis were robust,[12] the result is of little, if any, real significance. Most important, this "match" was something quite different from a "DNA match" as we know it in lay terms from paternity tests and *CSI* cases. Nevertheless, the news of the "match" sent a shock wave of ecstatic exclamations through the British media. "Yaba Daba Doo! Genes Prove Teacher's Old Man's a Caveman," wrote the *Mirror*, in a headline evoking the TV cartoon character, Fred Flintstone. The *Scotsman* described Cheddar Man as Targett's "long-lost relative," and the *Guardian* wrote about them as "close relatives."[13]

The media frenzy continued the following year, when a team of forensic scientists at Manchester University presented a facial reconstruction of Cheddar Man: a bust model featuring tangled dark hair, pale and puffy skin with a nasty boil above the right eyebrow, and hazel-brown eyes staring emptily into space.[14] The newspapers went into spin again, now with judgments on the looks of the Cheddar Man model and to what extent it resembled Targett and the current population of Somerset. Much interest was focused on the "blobby" shape of the nose, and Targett's wife Catherine entered the debate in defense of her husband's physical features: "Nobody could say Adrian has a blobby nose or a lopsided head," she said. "He has a nice nose."[15]

Brexit, Genomics, and the Importance of Color

Two decades then passed without much mention of Cheddar Man, before he was suddenly back on the front pages. This time, the people of the United Kingdom found themselves deeply divided over Brexit

(Britain's withdrawal from the European Union, which came into force in May 2021),[16] and in the years leading up to the split fierce debates raged over the essence of British identity. Were the British people an ethnically diverse and welcoming mirror to the world (as the British Museum is often described as a museum of the entire world), or was British identity instead defined by the prehistoric and precolonial roots of the nation, which needed seclusion and protection from the rest of the world to thrive? In these times of polarization and volatile political debate, the television network Channel 4 produced a documentary with the title *The First Brit*, which aired in February 2018. The opening scene features the British Union Jack flag waving alone in the sky to music of anticipation. A sonorous voice—familiar to fans of the drama series *Downton Abbey* as that of the rock-steady butler Mr. Carson—breaks in:

> There has been a lot of talk lately, about Britain—about who belongs, and who doesn't. Now, *science* is about to reveal the *truth* about where we come from, and who we *really* are. [. . .] Scientists will analyze the *entire* DNA of Britain's oldest complete skeleton.[17]

Cheddar Man was once again in the limelight, this time against a backdrop of Brexit polarization. The technical advancements of high-throughput sequencing were promising genome-wide analyses of aDNA, and there were resounding announcements of an "ancient DNA revolution." Archaeologists and geneticists at the Natural History Museum and University College London participated in the archaeo-science drama orchestrated by Channel 4 and were featured as they performed and commented on a new DNA analysis of the Cheddar Man skeleton. This time the sample was taken from the high-density petrous bone of the inner ear,[18] and unlike Sykes's earlier study of mtDNA, the analysis was now done across the whole range of the genome. But there were also similarities in the framing of the two studies.

First and foremost, they were both motivated by the idea that there would be a significant relation between Cheddar Man as the "first Englishman" and the current definition of British identity. Both Sykes's book and the Channel 4 production also put intense focus on the sampling procedure. It was described in words and images as expensive and high-stakes research, taking place in the secluded and secure spaces of advanced laboratories. "The ambitious project is *costly*, and the team knows the stakes are high," says the narrator in *The First*

Brit as the camera flicks between scenes inside the laboratory. A scientist in a white coat passes with a tray of test tubes, while another places a piece of bone under a microscope. Then comes a closeup on the microscope with the bone specimen, a clean workbench, a box of vinyl gloves, and someone in white protective clothing putting gloves on already-gloved hands.[19] The geneticists in charge comment on the pressure, and excitement, they feel as they are about to drill a hole in Cheddar Man's skull.

In both stories, the white bone powder that comes out of the tiny drilling hole is described as a precious treasure with nearly magic potential. Sykes describes the "slight smell of burning" at the vaporization of collagen as something he had "come to love as a sign that there was plenty of protein left in the specimen."[20] And in *The First Brit*, the voice of the narrator resounds as the white powder materializes in a magnified closeup of the drilling procedure: "These few milligrams of bone powder could contain secrets hidden for ten thousand years—where Cheddar Man came from, how he relates to us, and *exactly* what he looked like."[21] What happens next—in both stories—is quite remarkable. From the extracted bone powder there is a giant leap in the research procedure, all the way to the results. It is literally depicted as if the results come straight out of the white powder.

In *The First Brit*, the results are focused on the appearance of Cheddar Man, and the query of his "exact" looks is soon boiled down to a question of hair, eyes, and skin. The answer, which is revealed in the scene with the archaeologist and the computer scientist sitting on the stairs at University College London, is that Cheddar Man had "wavy" hair, "blue" eyes, and "dark-to-black" skin. These results would also be guiding the work with a new facial reconstruction.

A New Model

The new face model was made by twin brothers Adrie and Alfons Kennis, Dutch model-making artists known for the distinctive "personal character" in each of their models of hominins and famous "paleopersonalities" like Lucy and Ötzi. Their reconstruction of Cheddar Man's face departed from a 3D-scanned replica of the skull in London, and just like their other models it was meticulously researched on a number of details, such as the thickness of the forehead tissue for an average person of that age and gender. In their work with Cheddar Man, the Kennis brothers also had the benefit of the additional information from the DNA analysis, which suggested "dark-to-black" skin, "blue" eyes, and "wavy" hair.

If we stop for a moment and home in on the analyses between bone powder and presented results, we see that they are in fact quite complicated and involve a number of caveats.[22] One potential pitfall is that physical characteristics such as hair quality and skin color are decided by intricate interplays between genes in a number of different locations of the genome (of which the estimated effects are calculated and summarized as "polygenic scores") as well as environmental factors in the individual's lifetime. This means that the relation between genotype (the order of nucleotides in the DNA) and phenotype (the actual observable outcome) is neither straightforward nor easily predictable when it comes to these characteristics.[23] Skin color is particularly difficult to predict with precision, even in forensic cases where the genome is less fragmentary and where there is a living person with whom to compare the results.[24] This put a heavy responsibility on the Kennis brothers to interpret the meaning of "dark-to-black," "blue," and "wavy" into a true-to-life representation of Cheddar Man.

And this was not the only creative interpretation involved in the rendering of Cheddar Man as a person of flesh and blood. As an echo resounding from the Human Genome Diversity Project, the Kennis brothers have said that their works are inspired by "modern tribal peoples" in "isolated or primitive societies," which they study and use as "a window into the past."[25] The skull shape is thus often the only physical link to the featured individual. In the case of Cheddar Man, the museum used a high-tech 3D scanner, designed for the International Space Station, to capture the skull in full detail, and sent it to be printed as an exact replica at the Kennises' art studio in the Netherlands. However, state-of-the-art technology could never make up for the fact that the skull was found "much mutilated" when it was removed from the cave in 1903, so most of the front of the face was lost. Moreover, as explained by Adrie Kennis, "there are some things the skull can't tell you. You never know how much fat someone had around their eyes, or the thickness of the lips, or the exact position and shape of the nostrils."[26] And this information cannot come from DNA either, as far as it goes today. So even the most high-tech apparatuses and the most advanced statistical calculations cannot escape the fact that facial reconstruction is a creative enterprise. It is essentially a work of art.

The new Cheddar Man model (fig. 10) was unveiled at the Natural History Museum in February 2018 in the presence of the Kennis brothers and museum scientists involved in the study. At the unveiling, which is featured in *The First Brit*, a sigh of excitement is heard as the cover drops. The new Cheddar Man bust has dark and tousled hair with bangs—a bit like the earlier Manchester model, and not unlike the hairstyle of the artist brothers themselves—and bears the signature cheeky look of a Kennis model. But the star turn of the event is undoubtedly

Figure 10. Cheddar Man's face reconstruction at the Natural History Museum in London, August 2023. Photo by author.

the color of the eyes and the skin. In front of the camera, a seemingly spellbound archaeologist comments on the blue eyes and how they shine in the spotlight. The artist responds with a laugh: "Yes, they are the biggest eyes I have ever made!" The archaeologist continues:

> That combination of quite dark skin and blue eyes is something that we don't imagine is typical. But that was the real appearance of these people, something that is quite rare today. And we are not just conjuring this up out of nowhere. We really do have scientific data.[27]

But is this really true? In the details of the scientific report we find, in fact, quite a lot of ambiguity regarding Cheddar Man's pigmentation.[28] Due to fragmentation of the ancient genome, the analysis had to involve elements of guesswork. Moreover, as noted above, any prediction of pigmentation—even on living individuals—will be based on estimates of probability. The estimation of eye color is the most secure in terms of relation between genotype and phenotype, so it could be concluded with some confidence that Cheddar Man's eyes should have been somewhere between blue and green, possibly with spots of hazel. But importantly, the phenotype experts added that according to their analysis Cheddar Man would *not* have been a *"clear blue-eyed* individual." As for the more unpredictable skin color, three of the gene sections that are known to be involved in pigmentation could not be covered at all in the analysis, and two more had only been sequenced with low coverage, so these five sections had to be filled in by means of comparison with other genomes. This obviously affected the prediction model so that it did "not perform optimally." With these precautions clearly stated, the phenotype experts proposed "a dark complexion individual over an intermediate/light," and added that "it is unlikely that this individual has the darkest possible pigmentation, but it cannot be ruled out."[29]

All these details and ambiguities were lost in the translation from scientific report to featured model. It was even the case that some of the results stated in the report were ignored when the eyes were described as bright blue (although the model's eyes do, on closer inspection, include elements of hazel) and the dark complexion was amplified for maximum effect in the desired story of a "first Brit" with dark skin and blue eyes.

The Symbolic Work of DNA

We see here how the powerful *symbol* of new aDNA technology—with its aura of exactness and totality, its clean laboratories and costly machinery—gave almost limitless authority to the Kennis brothers' artistic model of Cheddar Man as a person of flesh and blood. And we see how the media team of *The First Brit*, the Natural History Museum, and the involved archaeologists and geneticists were all to some degree complicit in the making of the story of Cheddar Man as "Mesolithic Britain's blue-eyed boy."[30] As the new model is unveiled on camera, there is not a single question mark in sight, only steadfast statements referring to scientific data. All elements of estimation, guessing, interpretation, amplification, and creative construction are forgotten as Cheddar Man appears anew before the eyes of the museum scientists: "He is alive," says one of them. "He is a *person* now, he is not just—bones."[31]

The news about the dark-skinned Cheddar Man traveled fast and made headlines all over the world. The most featured message was a questioning of whiteness as a foundation for essential British identity, and it was in many cases presented as a long finger in the face of the most extreme Brexiteers. In this spirit, the *Guardian* sent a reporter to the village of Cheddar to interview the owner of a local pub and students at Adrian Targett's old school. To the students, who reportedly had "Pakistani, Irish, Greek Cypriot, Swedish, Sri Lankan, Guyanese, Indian and French roots," Cheddar Man's skin color sent a message of inclusion and tolerance. "The fact that he has the darker skin tone reminds us that we are all one race—the human race," said one of them. "That's what's most important. We should all respect one another." Another student drew more specific conclusions with regard to the current political situation: "With the rise of extremist groups like the English Defence League, people are getting more scared of immigrants. These groups tell people: 'You need to get out of the country because we were here first.' That's just factually wrong."[32] In another newspaper, Targett, who was now retired, commented that in this new version of Cheddar Man he could see a family resemblance in the shape of the nose and the blue eyes. As for his own personal identity, he added with a laugh: "I do feel a bit more multicultural now."[33]

Making Faces with aDNA

Some of the most spectacular uses of aDNA have been associated with the reconstruction of ancient individuals.[34] In the United States we have the Ancient One, also known as Kennewick Man,[35] and in Italy we have Ötzi the Iceman.[36] In Korea and Egypt, geneticists have published computer-simulated portraits of ancient individuals based solely on DNA,[37] and in 2019 a group of Israeli geneticists presented an artist portrait of a "Denisova girl" that gave rise to headlines like "Scientists Recreate the Face of a Denisovan Using DNA."[38] Just like Cheddar Man, these individuals are equipped with looks and characters that make sense in the particular contexts in which the reconstructions were created.

In all the cases mentioned above, aDNA studies mingle with the much longer tradition of facial reconstruction. Forensic anthropologist Catherine Wilkinson has defined facial reconstruction as "the scientific art of building the face onto the skull for the purposes of individual identification" and describes it as a tradition with deep historical roots that reflects "our unlimited fascination with human faces."[39] As such, the tradition of building faces onto prehistoric skulls has existed much more broadly and long before detailed aDNA studies were possible, but the new technology has brought a change in attitude to the face models. It has steered attention away from the professional and artistic skills of the model makers, to the presentation of the model as an immediate and animate image of the prehistoric person. As we have seen exemplified in the story of Cheddar Man, the reference to DNA tends to inspire a *Jurassic Park*–like notion of the real prehistoric individual "coming to life" in the reconstructed model. And as we have seen in the Cheddar Man case, this has little to do with the actual results of the DNA analysis. In this particular case, it could indicate and estimate some physical features, such as hair quality and pigmentation. So, the "coming to life" of the ancient person is a figment of imagination that falls back on popular images of genetic science where DNA figures as an exact blueprint of a person or, in the words of archaeologist John Robb, "as an oracle of identity."[40]

In the more recent cases from Korea and Egypt, as well as in the 2019 reconstruction of a "Denisova girl," DNA has come to play a decisive role as archaeogeneticists have set out to simulate faces of prehistoric individuals without having real skulls to depart from. Most extreme in terms of transfer from micro- to macro-level is the Denisova study, in which DNA extracted from minuscule physical remains (a third of a pinky finger from one individual) was used to make calculations of the most likely facial dimensions of "Denisovans" as "an extinct group of humans." The actual study was a methodological experiment in which patterns of methylation (a chemical reaction associated with the DNA molecules) in DNA from the "Denisova" pinky were compared with methylation patterns in DNA from samples of Neanderthals, modern humans, and chimpanzees. The results presented in the scientific paper are focused on the viability of the method of using "methylation maps" to predict anatomical features, and although the discussion includes a tentative prediction of Denisovan anatomy, it makes quite clear that the result is by no means conclusive.[41] Moreover, since there are no larger pieces of "Denisova" skeletons to compare with, the suggested anatomical features were deduced from similarities in methylation patterns with known features from Neanderthals, modern humans, and chimpanzees, and are described in broad comparative terms such as "robust jaws," "low forehead," and "enlarged mandibular condyle."[42]

These experimental and tentative predictions of anatomical features were then given to an artist who translated them into a painted portrait. Even more than the Kennis brothers in the case of Cheddar Man, the artist Maayan Harel had to use a great deal of imagination to turn generic anatomical terms into a painted portrait of "the Denisova girl." The result is an image of an apprehensive-looking brown-eyed girl with a prominent nose, slight chin with underbite, rough and dirty skin, and tousled brown hair. It is in many ways a classic popular image of a caveman,[43] albeit in the shape of a girl, and its relation to the scientific study is not entirely clear. Nonetheless, the image spread widely and was featured in news media and high-profile journals like *Science* and *Nature*, where it was complemented with headlines like "Ancient DNA Puts a Face on the Mysterious Denisovans" and "Denisovan Portrait Drawn from DNA"[44] Again, we note the symbolic power of DNA, even in serious science journals, as it offers credibility to the artist's portrait in spite of the unclear relation to the actual study.

Predicaments of Phenotyping

The use of aDNA for facial reconstruction also recalls critical discussions surrounding forensic DNA phenotyping—the practice of using DNA to construct images of suspects or victims of crime. To an outsider it may seem as if the image magically springs out of the DNA code, but the analysis is almost as complicated as in our examples with aDNA. Much like the Cheddar Man case, the DNA analyses in forensic phenotyping focus on specific areas of the individual genome that are known to affect physical traits, such as hair quality, eye color, and skin pigmentation, and by comparison with other known cases phenotype experts make predictions of possible physical traits for the suspect or victim.[45] Moreover, similar to the study of the Denisova girl, they make population-genomic estimations of ethnicity (sometimes referred to as "ancestry"), which are then used as generic templates for the drawings of individual faces.

This practice has been subject to critique because of its convoluted relation to the concept of race.[46] Anthropologist Roos Hopman has captured the fundamental problem eloquently in an anecdote from a coffee break at a conference for forensic phenotyping. Introducing herself as a social scientist interested in the use of race in phenotyping, she was interrupted by a surprised computer scientist who felt obliged to lecture her that "biological race does not exist" when it comes to humans, and who added that "modern methods are moving further and further away from race." A little later in the conversation, the same computer scientist commented on a facial reconstruction he worked with:

> The problem with this reconstruction, he explained, was that the skull (of which only the upper cranium was intact) had been given a lower jaw that was "too Western European." He had solved the problem, he went on, by reconstructing a "more Mongoloid" jaw.[47]

Revealed here is a paradox that is fundamental to studies of aDNA as much as it is to forensic phenotyping. Ever since the issuing of the oft-cited statement by UNESCO that "scientists have reached general agreement in recognizing that mankind is one: that all men belong to the same species, *homo sapiens*," the nonexistence of biological human races has become something of a truism in the world of science and academia.[48] Moreover, in the wake of the Human Genome Project—not least in the context of the "aDNA revolution"—it is often heard that the new big-data technology for analyzing DNA is moving us steadily and safely away from old, murky race science.[49] It has also been said that the expanded knowledge of polygenic traits or diseases (meaning that multiple genes in different parts of the genome are involved in producing a single trait or characteristic) makes current genetic science immune to ideas of racism or policies of eugenics.[50] Yet, as Hopman and others have shown beyond doubt, old systems of racial classification tend to be "folded" into new genomic practices.[51] In the conference anecdote, the computer scientist resorted to crude terms such as "Mongoloid"—which derives from nineteenth-century race science and is now considered "highly offensive" due to its association with "old racial (and racist) theories"[52]—to describe facial morphology of now-living people. It seems the truism of the nonexistence of human races and the mantras of new genomic science have made some scientists so secure and confident of their own antiracist position that they see no need to be vigilant of tendencies to racialized thinking in their own practice.

We have no reason to believe that this is done with bad intent. But this kind of ignorant confidence is dangerous, especially if it leads to the development of new forms of race science performed under the banner of antiracism.[53] For even if we avoid stark and crude terminology from early race science, we cannot escape the fact that current technologies for DNA phenotyping depend on the comparison of individual genomes with "populations" that function as generic racial, national, or ethnic types. In phenotype-inspired portraits or face models of prehistoric individuals, therefore, notions of race differentiation and essential ethnic identity are always lurking.

On a sunny Sunday in August 2023, I went to see the display of Cheddar Man at the Natural History Museum in London. I was there with my son and his friend, and we did several rounds through the crowded labyrinths of exhibitions before we eventually found it, hidden away in a cul-de-sac of the department of human evolution. "Wow! Look at that weenie!" one of the boys burst out as we passed the Kennis model of a Neanderthal man. The display of Cheddar Man was tucked away, off the main street of the exhibition, and was more modest than I had expected. Along with the skeleton, which was laid out in supine position, there were a couple of short texts with the titles "Where did we come from?" "Cheddar Man," and "Modern Humans in Britain." And then there was the Kennis reconstruction, encased behind glass (fig. 10) with the text:

> This is the reconstructed head of Cheddar Man based on the shape of his skull and DNA analysis. DNA research carried out with scientists from the Natural History Museum revealed that he was blue-eyed and had dark skin pigmentation. DNA was also able to confirm that he was male, which was suspected because of the shape of his skull and pelvis. Scientists think that the hunter-gatherer population Cheddar Man belonged to died out as farming communities spread across Britain.

As expected, the DNA analysis assumes prime position as a way toward revelation and confirmation. But the sentence at the end is noteworthy: Scientists think that "the population" he "belonged to" had "died out." How can we understand this, especially in light of the media frenzy around color diversity and multiculturalism as the proper foundation of the British nation? Moreover, can we really say that Cheddar Man "belonged to" a population (in the past tense), when that population was in fact constructed in twenty-first-century computers using modern models of population genomics? How likely is it that the Cheddar Man individual's sense of belonging back in prehistoric days was a match with those models? And finally, can we say that a population has "died out" while maintaining that the science has nothing to do with the construction of race or racial typology? If we are all related, if humanity is one single family, and if ancestry is something complicated and very messy—then why do we still find the need to talk about genetic populations as essential, and possibly extinct, historical communities?

The Making of Paleopersonalities

At first glance, the Cheddar Man models and the creative portrait of the Denisova girl seem like innocent and playful attempts to bring the long dead back to life. But they deserve our critical attention because such models and images are power tools in scientific knowledge production. As noted by Marianne Sommer, they are "no mere ornaments of texts, they constitute theories and contain elaborate arguments, feeding back into the scientific debate."[54] The media hype around Cheddar Man, moreover, indicates that stories and images of ancient personalities, especially when leveraged by the "oracle" power of DNA, can stir affective responses that have considerable effect on public opinion and political debate.

Another ancient individual who achieved fame via DNA and stirred affective responses is the "Female Viking Warrior" from Birka. The hype started in September 2017 with the publication of a scientific paper that reported a DNA study of a skeleton found in 1878, in a grave with two horses and typical warrior equipment, in the Swedish Viking Age town of Birka.[55] The bones from the Bj 581 grave had then been stored in the collections of the Swedish History Museum in Stockholm. At some point the skull had gone missing, but the shape of the pelvis clearly indicated that the skeleton was that of a biological female. However, despite unanimous confirmation from two osteological studies (the first in the 1970s), parts of the archaeological community had been reluctant to accept the idea of a woman buried as a Viking warrior. This spurred a local team of archaeologists and geneticists to perform a DNA analysis on the skeleton, and, so to speak, settle the issue for good.

Chromosomal sex determination is relatively simple to do, even on fragmentary aDNA, and does not require a genome-wide analysis. Nonetheless, a full genome analysis was done, and the unsurprising results were published in a paper with the title "A Female Viking Warrior Confirmed by Genomics." Within hours, the news about the female Viking warrior exploded into an international media nova,[56] with headlines such as "Famous Viking Warrior Was a Woman, DNA Reveals" and "DNA Proves Fearsome Viking Warrior Was a Woman."[57]

In the essay "The Lagertha Complex," my colleague Andreas Nyblom unpacks a fascinating process of personification, in which the anonymous skeleton from grave Bj 581 is endowed with a persona that has the looks of a Hollywood shero and the ambitions of a twenty-first-century postfeminist.[58] He demonstrates how this persona was born in an intricate interplay between academic researchers and popular media, most notably the production team of the TV series *Vikings*, that began long before the publication of the scientific article. Even more than the Cheddar Man case, we see here that DNA plays a merely symbolic role in a story that already existed long before the analysis. It also gets to metaphorically prove much more than the chromosomal sex of the individual in Bj 581, as DNA seems to bring this individual to life in a "fearsome" and "kick-ass" personification of the legendary shield-maiden Lagertha.

If the new Cheddar Man was focused on color, the Viking warrior was all about sex. Few media outlets took interest in the chromosomal sex of Cheddar Man, just as there was no mention of the eye and skin color of the Viking warrior. There were indeed other, genetic and strontium isotope, analyses done on the Birka remains, but none of that gained traction in the media hype. Only the sex was of interest—and then not in terms of chromosomal sex, but translated into a full-fledged twenty-first-century female gender identity, juxtaposed with the "Viking" warrior as a traditionally male-gendered character. This testifies to the importance of contemporary context. "Each generation projects onto the Neanderthals its own fears, culture, and sometimes even personal history," say Erik Trinkaus and Pat Shipman. "We do not see things the way they are, we see things the way we are."[59] Likewise, both Cheddar Man and the female Viking warrior are entwined in identity politics specific to the times and situations in which they were created as DNA-inspired personalities.[60]

The identity-politics importance of these personalities is evident in each case from the subsequent media hype, as members of the public got caught up in the stories of Cheddar Man and the female Viking warrior and related them to their own identities, and to political struggles and ideals in society at large. But we also note how the researchers in both cases fueled the identity-politics potential by expressing their own positive excitement. Interestingly, in both cases it is the archaeologists who are the most expressive. In *The First Brit*, the archaeologist says it is "brilliant" when he learns about Cheddar Man's pigmentation, and in a similar docudrama production about the female Viking warrior, the first author of the "Female Viking Warrior Confirmed by Genomics" paper says with confidence that the individual buried in Bj 581 "would have loved for us to be standing here today, talking about her, and talking about her achievements more than a thousand years later."[61]

Using People in the Past

Ever since those first pieces of a human skull were found in Gough's Cave in 1903, the story of Cheddar Man has moved to and fro between profitable farce and high-stakes identity politics. In the twenty-first century, we have seen the interests shift with new possibilities of genomic science—from the focus on ancestry and kinship that was allowed by analyses of mitochondrial DNA, to the recent genome-wide analyses that promise predictions of phenotypic expressions of physical characteristics such as color.[62] But at the end of the day, it is the stories that we remember, the wild kind of popular history-writing that is fueled by the use of aDNA—especially, it seems, in polarized political contexts where the academic community is more or less safely sorted on one side of the debate.

Just to make it clear: I too would much prefer a definition of current British (and Swedish) identity as diverse and tolerant. I also want schoolchildren (everywhere) to feel that we are all one race—the human race—and that we should respect one another. I would also, every day of the week, promote a view of gender that allows individuals of any chromosomal sex to pursue whatever professional goals they want. But is it right to use ancient individuals as props in that pursuit? Is there an ethical limit to our exploitation of the long dead, when we provide them with new forehead tissue and bright glass eyes, or claim that they would have been excited about the spotlight we have shone upon them?[63] Also, what are the risks of using DNA as a stamp of proof on these entire personality packages, which are so clearly chiseled out and spotlighted for our own purposes? How are we then going to argue with credibility against the use of DNA as proof in stories that we find dangerous, stories that may lead to persecution, violence, or even genocide? These are complicated questions, and I do not pretend to have an easy answer. But I will ask you to keep them in mind as we now move on to the final chapter.

FURTHER READING

Hopman, Roos. "The Face As Folded Object: Race and the Problems with 'Progress' in Forensic DNA Phenotyping." *Social Studies of Science* (2021).

Nyblom, Andreas. "The Lagertha Complex." In *Critical Perspectives on Ancient DNA*, ed. Daniel Strand, Anna Källén, and Charlotte Mulcare. Cambridge, MA: MIT Press, 2024.

Oikkonen, Venla. *Population Genetics and Belonging: A Cultural Analysis of Genetic Ancestry*. London: Palgrave Macmillan, 2018. Chapter 3.

5: In Defense
of the Molecule

"We are not finders of fact. We are tellers of story."

"You base the story on the evidence, no?"

"No! We base the *evidence* on the *story*. We prove what helps us, and we disprove what hurts us. Whoever tells the best story goes home with the cash-in prizes."[1]

In a third-season episode of the legal drama/political satire *The Good Fight*, maverick lawyer Roland Blum gives junior colleague Maia Rindell a lecture on how to survive and succeed in court. The series features a group of Chicago lawyers struggling with the polarized society and chaotic public discourse of the Trump era. Words such as "facts" and "evidence" are devalued, and the truth is up for grabs for anyone with enough power and platform to sell their story.

This has also posed challenges in the realm of scholarship. In a public discourse spun by the affective logic of social media algorithms—where any kind of extreme black-or-white message is rewarded, and sensible analyses in various nuances of beige are of little value—it seems more important than ever that science and academia rein back, honor the labor of serious analysis, and stay true to the (often beige) nuances of the real world. I also find reason to recall Donna Haraway's principles of *situated knowledge*.[2] A responsible science stays true to its object by acknowledging the situated perspective of any knowledge claim, and by accounting honestly and transparently for the machinery (intellectual and technical) involved in the analytical process.

Minding the Illusory Truth Effect

Contrary to the principles of situated knowledge, most of the stories I have presented in this book hinge on the illusory truth effect of aDNA as a petrified book of life or an oracle of identity.[3] I have demonstrated how the scientific mapping of relations of genetic material in the ancient past have inspired colorful and politically potent stories of real groups of people on the move across real landscapes. Genetic similarity calculated with computer statistics between a few samples across vast continents and spans of thousands of years have been compressed into compelling stories and images of actual journeys by actual people. Arrows have been used as impactful illustrations of the movements of these seemingly real groups of people as they appear to have migrated, traveled, or galloped with purpose and intent from old points of origin to new destinations. Family trees and PCA diagrams that depend on constructed computer models calculating genetic relations have portrayed humanity as essentially divided into separate bioethnic groups. Haplogroups and "matches" have been widely cited as proof of meaningful relations between people in the ancient past and the present. Genetic sex determinations and speculative phenotype estimations have been invoked as evidence for the truthfulness of artistic renderings of the looks and personalities of prehistoric individuals. At the foundation of these stories is the widespread notion that DNA has the power to reveal the *exact* identities of people in ancient times, as well as the full and final story of our relations to them. And this notion hinges on the illusory truth effect of DNA rather than on the real potential of the DNA molecule.

The illusory truth effect was first defined in 1977 by a group of American psychologists in a study of memory and learning. Exposing a group of college students to a set of true and false statements on issues they had no prior knowledge of (such as "Lithium is the lightest of all metals" and "Zachary Taylor was the first president to die in office"), repeating some of the statements after two weeks, and then again after another two weeks, the researchers could detect a clear trend: the students tended to regard statements as true if they had heard them many times, regardless of whether they were true or false.[4] These findings established the truth effect as a concept in research on memory and learning, but for some time it was assumed that it worked only in instances where the subject had no prior knowledge of the issue at hand. This assumption was challenged in a 2015 study, which exposed subjects to true and false statements on issues they were both familiar and unfamiliar with. The results indicated, contrary to earlier assumptions, that even if the subject had previous knowledge of the issue, they tended to accept a false statement as true if it was repeated enough times. Hence it could be concluded that repetition can trump rationality when people assess a statement as true or false.[5]

This is food for thought when we set out to write history with aDNA. It means that even if most people *know* that DNA is not a simple blueprint of life, or "a traveler from an antique land who lives within us all,"[6] they are likely to accept these fictions as true if they are repeated often enough. This gives us every reason to be concerned by the reckless kind of storytelling that has been associated with aDNA studies.

The Unfortunate Image of Pure Science

Considering the popular appeal of both DNA and archaeology, it should come as no surprise that aDNA research has been attuned to prospects of media attention. Historian of science Elizabeth Jones has studied this aspect of aDNA studies in terms of "celebrity science," and her interviews with practitioners in the field reveal an astute awareness of the newsworthiness of their research. One of her (anonymized) interviewees explains some of the allure of aDNA research when they describe it as appearing to give "access to what you might intuitively think is unreachable, unknown, and mysterious." In this way, the allure of ancient DNA makes the field attractive to popular media in search of sensational, salable stories. Another interviewee fills in: "We always have journalists ringing us and saying [. . .], 'I need a story for something. What have you got?'" And they further explain how they feed that interest:

> No one really wants to read about the peptidoglycan in bacteria cell walls. It might be very important—probably much more important [. . .]—but [. . .] your average person is not going to read that. But you can always write a good story about a king or a mammoth or whatever.[7]

Such quotes suggest that the story-writing element of aDNA research is something taken lightly by (at least some) practitioners in the field. It is here described as a lightweight enterprise that is shamelessly attuned to current media interests, that can be done with ease, alongside the "much more important" scientific work. In my experience of talking and listening to practitioners in the field, this attitude does indeed exist. It reflects a widespread image of aDNA studies as divided into two essentially separate enterprises: on the one hand, a serious scientific procedure performed with methodological rigor in clean, state-of-the-art laboratories, and, on the other, a lightweight, media-oriented pursuit that aims to satisfy eager journalists' need for a good story or the general public's hunger for exciting and easy-to-access knowledge about the ancient past. In this divided image of aDNA studies, the latter aspect has tended to be dismissed as unimportant "popularization" or "misuse" of scientific results.[8]

This is unfortunate, for several reasons. First of all, it is not an accurate depiction of the practice of aDNA research. Even the quickest glance at the practices within this field—as seen in various examples from the previous chapters—will reveal that knowledge of aDNA is formed in convoluted interactions between a complex set of materials and actors, inside and outside laboratories, in scientific and academic journals, and in popular media productions.[9] In reality, these actors are by no means siloed in the work they do in relation to aDNA research. On the contrary, they are essentially dependent on one another in what could be described as a neoliberal ecosystem of academic production.[10] A high-impact science journal could not survive long in our time without broad interest in the articles published. In many countries, not least in the United States, academic researchers depend on publications in these journals if they want to pursue successful careers. Moreover, with its bespoke laboratory facilities and big-data methodologies, aDNA research is quite a resource-intensive enterprise.[11] Researchers with grand ambitions must thus present and sell knowledge-claims on the same level as the hefty resources they need to pursue their goals, to motivate funding as well as to secure precious aDNA samples. As we have seen in previous chapters—from variations on the story of "who we are and how we got here" to the analyses that inspired the "real" looks of Cheddar Man—results are being forged, stretched, and spun into stories to meet grand expectations. This is no big revelation to anyone working within the field. It is common knowledge that in order to succeed, you need to play the game—and the game involves courting publishers and academic journals, popular media, and funding agencies just as much as working in the laboratory.

Yet, the resilient image of an enterprise essentially divided between serious science and lightweight popularization has come to serve the field and its practitioners in important ways. It is expressed in popular renderings of aDNA research that tend to revolve around pictures of white overalls in sci-fi-style laboratories, combined with suggestive imagery of "prehistoric" people in "primitive" clothing walking across vast, empty landscapes or meeting in violent or romantic encounters.[12] As in the example of Cheddar Man, the process of extracting DNA is commonly depicted as a rigorous procedure requiring highly specialized skills, protective clothing, and unique laboratory facilities.[13] In such renderings, aDNA extraction can easily appear to the outsider as a magic process mastered by wizards in white space-walk-style overalls. And the precious nature of samples, skills, and unique laboratory facilities conveys a sense of value and mystique to aDNA studies altogether. But when we focus solely on the extraction of DNA, we tend to overlook the hazards and fragilities involved in other parts of the research process. All other phases of the analysis—the all-important formulation of research questions, the selection and labeling of samples and reference databases, the selection of statistical models, the writing of algorithms, the use of diagrams and other visuals, and the choice of words and narratives to present the results—all of that which essentially decides the perimeters, structures, meanings, and overall impact of the results, is upstaged by images of white bone powder and protective overalls in clean laboratories. When combined with suggestive images of "prehistoric" people, it seems as if ancient individuals and groups of people appear neutrally out of bone powder and test tubes, when in fact their physical features and group formations are predicted to no small degree by the statistical models and visualization technologies chosen by the researchers.

The Illusion of Misuse

When the results of aDNA research have come to prove useful in inflammable social and political situations—when, for example, the neo-Nazi Internet forum Stormfront picked up the studies of Yamnaya migrations as scientific confirmation of the origin and male-warrior

character of Aryans,[14] or when Israeli Prime Minister Benjamin Net-
anyahu used a study of aDNA from an archaeological site in southern
Israel to fire off a provocative tweet about the Jewish people arriving
long before the Palestinians to the Holy Land[15]—the researchers responsible for the studies have chosen to withdraw and close the door
to the debate.

In interviews with journalist Howard Wolinsky, two of the researchers who were responsible for the genetic studies of Yamnaya and the
site in southern Israel explained how they see their own involvement in
these particular situations. The one responsible for the Israel study says
that he chose not to get engaged, although he saw the research results
being interpreted in wrongful and potentially harmful ways, because
it would "not be a scientific debate. [. . .] It's politicizing, and we have
been installed by the government to do the science that we do."[16] The
other, who was responsible for the Yamnaya study, says of the Israeli
debate that "citation of genetic research in this way to support territo-
rial claims is a truly sort of abhorrent and non-fact-based perspective
of the past," and says more generally about social media that "much of
it is extremely low-level material and not serious material," so he, too,
chose not to get engaged in the Yamnaya debates.[17] These reactions are
logical if we abide by an image of aDNA research as essentially divided
in two parts: serious neutral science, on the one hand, and unimportant
popularization or "misuse," on the other.[18] But this image is an illusion.

A similar image occurred in the debates surrounding the Human
Genome Diversity Project in the 1990s. When the HGDP was accused
of being exploitative and encouraging racist stereotypes, its proponents started to label the negative consequences of their research and
popular communication in terms of "misuse." It was not a sustainable
image there either, as has been argued by sociologist Jenny Reardon
and more recently by philosopher Natan Elgabsi.[19] In reality, genetic
science cannot be separated from its meaning-making context. It was
the meaningful messages of groups of people on the move through
history, and the promises to explain who we are, where we came from,
and how we are all related, that gave this research field the funding and
visibility on which it continues to thrive. The same messages that are
deemed "abhorrent" and classified as "misuse" when they are activated
in dangerous political contexts are thus in fact the raison d'être for this
kind of research.

Illusion or not, the image of an essentially divided enterprise has been fundamental to the developments of ancient DNA research. It has given some researchers leverage and leeway to step into the limelight and present stories that will benefit them—whether through access to samples, the opportunity to publish in a high-ranking journal, funding, acclaim, or the satisfaction of "proving" a story that they always wanted to tell—and then retreat back into the black box of the clean laboratory as soon as a storm rises at the horizon.

A Call for Transparency

The DNA molecule has no bad intention. It plays an important role in all forms of life, and the way it is copied to generation after generation makes it a suitable subject of research on biological family relations. From that perspective, we have every reason to welcome aDNA research in studies of prehistory.

But let us not be naive or ignorant. If we want to write history with aDNA, we must first acknowledge how knowledge has been structured in the history of aDNA research. It has always leaned heavily on the technologies and traditions of population genetics. This will make aDNA research inherently useful for various political projects that take interest in essential identity and exclusive groups of people moving steadily up or down through history.[20] When the results are picked up in ethnonationalist contexts, it should not be dismissed as occasional misuse of neutral research. It should rather be expected, since it is inscribed in the very structure of population genetics when applied to prehistoric or historic research.[21] With the great symbolic value invested in DNA as evidence, the stories that are presented as results of aDNA research tend to outweigh other history-writing and can thus have more serious consequences. Moreover, aDNA research is particularly vulnerable because dead people cannot say no. Unlike the critics of the Human Genome Diversity Project, ancient people cannot object to the categories and characteristics that are ascribed to them.

For all those reasons, it should be in the interest of every responsible scientist and scholar working with aDNA to be open and transparent about the many caveats and vulnerabilities of this kind of research—not only in outlining their methodologies in scientific papers (which for nonspecialists are black boxes with very thick walls), but also up front in more readily accessible article abstracts, interviews, and public communication. I take my hat off to forensic geneticist Susan Walsh, who stood up in the midst of the Cheddar Man media frenzy and explained that her research results—results that had just inspired a tsunami of headlines and exclamation marks in popular media around the world—were based on statistical modeling of probabilities with plenty of caveats and should not be read as an absolute prediction of Cheddar Man's real skin color. "It's not a simple statement of 'this person was dark-skinned,'" she said in an interview. "It is his most probable profile, based on current research."[22]

Let us take inspiration from Walsh and strive for more transparent communication.[23] Let us resist the temptation to use the term "whole genome sequencing" as if it meant an actual sequencing of all the 3 billion base pairs of an ancient human genome, when in fact we work with samples in which only minuscule bits are preserved and we have to create computer models to simulate the rest.[24] Let us not use images of clean, state-of-the-art laboratories and powerful computers as pretext for a view of genetic science as a neutral machine where unbiased truths about ancient people's identities come straight out of bone powder. Let us instead speak openly about the caveats of aDNA sampling when only a few individual genomes among millions are available for analysis, and explain how we *construct* groups of people in ancient times by means of mathematical models, applications of Occam's razor, and tantalizing visualizations. Let us be clear that "Portuguese" is not the same as Portuguese, and that the Denisova hominin was not found or discovered as "an extinct group of humans," but was rather *invented* as a named collective to account for a probability induced by the statistical models of population genomics. Finally, let us not exploit the allure of "the DNA match," but instead be open and clear about *what*, exactly, it is that is matched.

In Defense of the Molecule

This book is in defense of the DNA molecule. Molecules do not speak, and should not be burdened with responsibility for stories they have not told. If anything, as noted by Jerome de Groot with reference to the poet Michael Symons Roberts, "DNA defies our attempts to impose meaning and to 'read' and 'see' it."[25] If we decide to take full responsibility for our own stories—not pretend that they appear as by magic out of the molecules, and not exploit the illusory truth effect of DNA—the field of archaeogenetics will undoubtedly lose some of its current benefits. The allure and mystique that have so far driven much of the interest in the field would be unveiled, exposing an emperor sparsely clothed.[26] But naked is not necessarily bad. It could be our best way to find relief and emancipation in a time characterized by identitarianism, extreme story spinning, and alternative truths. If we stay as close and true as we can to the naked molecule in studies of aDNA, we will be better equipped when the storm rises and we might need to explain that our research does not support political projects of persecution, violence, or even genocide.

Indeed, there are many existing studies of aDNA that have been done with a closer focus on the molecules, without grand ambitions to sell mind-blowing stories about the history of everyone and everything in all times.[27] In this book I have written about human DNA, but there are also studies of aDNA from animals, plants, pathogens and other microorganisms[28] that have offered important insights for prehistoric and historic research without ambitions of grand storytelling.[29] In research on human aDNA, there have been calls for more intensive, fine-scale studies of a larger number of individuals in a limited context, such as an ancient cemetery.[30]

Genetic analyses targeting biological family relations, chromosomal sex, and phenotype estimations can in such studies complement archaeological interpretations of material culture and burial contexts, isotope analyses indicating lifetime mobility and diet, and osteological analyses of age, pathology, and physical attributes in a nuanced and fine-grained analysis of ancient individuals and communities. Existing studies along these lines remain weighted in favor of genetics,[31] suggesting there is still some way to go to achieve a fully balanced interdisciplinary collaboration.[32] But there is promise in the intent to scale down to a finer grain, assume a more humble position, and allow other perspectives on life and identity to complement and enter into balanced conversation with the findings of genetic science.

But we must remain aware that, even when we do stay close to the naked molecule, we will never be entirely free from the structure of knowledge and definitions of identity inherent in genetic science. Nor will we be able to escape the illusory truth effect of DNA, with tantalizing images already seen and ambitious stories already heard. Again, let us listen to Donna Haraway, here in a passage inspired by anthropologist Marilyn Strathern:

> It matters what matters we use to think other matters with; it matters what stories we tell to tell other stories with; it matters what knots knot knots, what thoughts think thoughts, what descriptions describe descriptions, what ties tie ties. It matters what stories make worlds, what worlds make stories.[33]

In other words, we can never escape the many contingencies of the aDNA molecule in current discourses if we want to say something meaningful about it. It will take hard work to navigate the troubled waters of already public knowledge. The best we can do is to "stay with the trouble" and embrace the complexity in honest and transparent conversations. It could be described as an opposite strategy to Occam's razor. In honor and acknowledgment of the always complicated, always situated nature of human identity, we ought not to rush to the simplest explanation, but should always be prepared to look for the more complex picture.[34]

Holding Both

A possible way forward is presented to us by cultural theorist Trinh T. Minh-ha. Many of her works have revolved around the predicaments of binary thinking, and she has recently called for a new, nonbinary figure of thinking and speaking in which we strive to "hold both," embracing two sides at once. "Low and high, old and new, north and south" can be regarded as "non-binary twos," she says.[35] The history of genetics has thus far been characterized by the binary tension between hope and fear. Considerable efforts have been put into separating the two as far as possible, by convincing the public to invest in the hope that genetics will contribute to a better future, and not to dwell on the fear of potentially dangerous consequences.[36] Perhaps it is now time that we learn to hold both.

We should not expect it to be a quick fix or an easy task. If we are serious in our effort to hold both, it will not be enough to make critically aware statements in ethical guidelines and best practice articles, and then continue with business as usual. It will require a lot of labor—intellectual, social, emotional—to make two sides of a comfortably contrasted figure merge into a nonbinary two.[37] It means that we need to take in, and take responsibility for, the fact that there will always be a dark side to the promises of genetic science. Genetic projects with antiracist ambitions can be compatible with racism. The positive potentials of ascribing identities to people in the past and the present with the authority of genetic science will always come with a possibility for dangerous political use. Humans, now as in ancient times, are cultural *and* biological beings. Ancient DNA is a material *and* an imagination. When we write history with ancient DNA, we are finders of fact *and* tellers of stories.

Acknowledgments

This book is based on the knowledge I have gained from working with Charlotte Mulcare, Andreas Nyblom, and Daniel Strand. Owing to their loyalty, resilience, and humor, we stayed afloat during the extraordinary times of the COVID-19 pandemic. From the bottom of my heart: Charlie, Andreas, and Daniel—thank you for all the laughs, the brilliant ideas, the sometimes trying but always rewarding processes of writing and cowriting, the eye-opening disagreements, and the lightbulb moments we shared in the Code Narrative History project. Much of the contents of this book are yours as well as mine. The well-worn words "it could not have been done without you" have never been said more truly.

I also want to extend my gratitude to Riksbankens Jubileumsfond—the Swedish Foundation for Humanities and Social Sciences—for supporting our project with a generous research grant, and to Stockholm University and the Department of Culture and Aesthetics for hosting it. Bérénice Bellina, Andreas Gunnarsson, Erika Hagelberg, Charlotte Hedenstierna Jonsson, Anna Kjellström, Alison Klevnäs, Per Larsson, Ian Lilley, Jonathan Lindström, Christine Lorre, Lynn Meskell, Elisabeth Niklasson, Christopher Prescott, Mélanie Pruvost, Oli Pryce, and Howard Williams have supported and enabled our work with crucial inputs. Faculty, students, and members of the research seminars at Stockholm University, the Department of Archaeology and Classical Studies, and the Department of Culture and Aesthetics; at Uppsala University, Centre for Multidisciplinary Studies on Racism; and at Umeå University, Department of Culture and Media Studies, have offered continuous support and inputs to our research. Presentations of our results in these contexts, as well as at the University of Aberdeen, Cornell

University, Stanford University, and the University of Oslo, have been important stations on the way to completing this book. My warmest thanks to all who attended these talks and seminars and contributed with information, questions, and criticisms, which have broadened my perspectives and added to my knowledge.

I am moreover grateful to colleagues and friends who contributed to our special issue of *Journal of Social Archaeology* (21, no. 2 [2021]) and our edited volume, *Critical Perspectives on Ancient DNA* (Cambridge, MA: MIT Press, 2024). Ruth Amstutz, Deborah Bolnick, Elsbeth Bösl, Chip Colwell, Amanda Cortez, Magnus Fiskesjö, John Hawks, K. Ann Horsburgh, Judith Jesch, Elizabeth Jones, Stewart Koyiyumptewa, Amade M'charek, Venla Oikkonen, Mélanie Pruvost, and Marianne Sommer have contributed with multidisciplinary expertise and bright minds to the pursuit of a more profound understanding of the current developments in ancient DNA research. Our conversations during workshops, editorial meetings, and text editing have been instrumental for my own knowledge pursuits and are fundamental to the ideas and conclusions presented here.

Crucial inputs also came from conversations with members of research teams working with ancient DNA in Sweden and France, and interviews with archaeologists, curators, and science journalists across the world in preparation for our *Current Anthropology* article "Petrous Fever: The Gap between Ideal and Actual Practice in Ancient DNA Research" (2024). These informants have to remain anonymous, but the critical astuteness and analytical qualities of the information they shared have been absolutely instrumental for our research in the Code Narrative History project, and for the arguments presented in this book. For that I am ever grateful.

The writing of the book has been enabled and supported by the stellar editorial team at the University of Chicago Press. A special thanks to Karen Levine, Victoria Barry, and Stephen Twilley, and to Lys Weiss of Post Hoc Academic Publishing Services, for letting me benefit from their professional expertise, putting pressure where needed, and adding crucial tweaks that turned a first manuscript into a complete book. I am grateful to David Robertson for preparing the index. A very warm thanks also to Marianne Laimer, for the skillful drawing of the map and dendrogram in figures 4 and 6.

Along with the members of the Code Narrative History team and three anonymous reviewers, Victoria Fareld, Johan Hegardt, Adam Hjorthén, Andrew Jones, and Elisabeth Niklasson have read and commented on an earlier draft of the text. Their sharp-eyed indications of gaps and inconsistencies and their generous suggestions for improvement have been invaluable for the development of the manuscript from draft to finished book. I am most grateful for their contributions and am alone responsible for any remaining errors.

Last but not least, the most important team, which makes it all worthwhile. Otto, Hedvig, Hugo, Per Arne, Marianne, Jonas, Maja, and Torun—thank you for always being there with constant practical and spiritual support, and as a beautiful reminder of the great wonders of family relations, with or without genetic kinship. Johan—this book would not have existed without you. You have given me courage and essential writing time by cooking, listening, cheering, cooking, picking me up when in doubt, listening, adding important details to my arguments, reading, cheering . . . and cooking. Thank you so much.

Notes

Introduction

1. Hanna Rydh, *Bland fornminnen och indianer* (Stockholm: Natur och kultur, 1934), 135. In the Swedish original: "Sysslandet med förhistoriska epoker ger vetenskapsmannen många tillfällen att inse sin egen ringhet och oförmåga och detta på mer än ett sätt." My translation.
2. Michael Price, "DNA Proves Fearsome Viking Warrior Was a Woman," *Science*, September 8, 2017.
3. Cheyenne MacDonald, "First Proof There Really Were FEMALE Viking Warriors: DNA Study Reveals High-ranking 'Valkyrie' Buried with Weapons and Two Horses Was a Five Foot Six WOMAN," *Daily Mail*, September 11, 2017.
4. Charlotte Hedenstierna-Jonson et al., "A Female Viking Warrior Confirmed by Genomics," *American Journal of Physical Anthropology* 164 (2017): 853–60.
5. For further details, see Andreas Nyblom, "The Lagertha Complex: Archaeogenomics and the Viking Stage," in *Critical Perspectives on Ancient DNA*, ed. Daniel Strand, Anna Källén, and Charlotte Mulcare (Cambridge, MA: MIT Press, 2024).
6. Some other analyses were done, but none of them gained traction in the media reports.
7. Anna Källén et al., "Archaeogenetics in Popular Media: Contemporary Implications of Ancient DNA," *Current Swedish Archaeology* 27 (2019).
8. David Reich, "How Genetics Is Changing Our Understanding of 'Race,'" *New York Times*, March 23, 2018.
9. Jonathan Kahn et al., "How Not to Talk about Race and Genetics," *Buzzfeed*, March 30, 2018.
10. Reich, "How Genetics Is Changing Our Understanding of 'Race.'"
11. Among their many works, these have been the most fundamental to my own thinking: Donna Haraway, "Situated Knowledges: The Science Question in Feminism and the Privilege of Partial Perspective," *Feminist Studies* 14, no. 3 (Autumn 1988): 575–99; Donna Haraway, *Primate Visions: Gender, Race, and Nature in the World of Modern Science* (New York: Routledge, 1989); James Clifford, *The Predicament of Culture: Twentieth-Century Ethnography, Literature, and*

Art (Cambridge, MA: Harvard University Press, 1988); James Clifford, *Routes: Travel and Translation in the Late Twentieth Century* (Cambridge, MA: Harvard University Press, 1997); Bruno Latour, *Science in Action: How to Follow Scientists and Engineers through Society* (Cambridge, MA: Harvard University Press, 1987); Michel Serres and Bruno Latour, *Conversations on Science, Culture, and Time* (Ann Arbor: University of Michigan Press, 1995).

12. For example, see Trinh T. Minh-ha, *Woman, Native, Other: Writing Postcoloniality and Feminism* (Bloomington: Indiana University Press, 1989); Trinh T. Minh-ha, *Elsewhere, within Here: Immigration, Refugeeism and the Boundary Event* (New York: Routledge, 2011); Edward Said, *Orientalism* (New York: Pantheon, 1978); Homi K. Bhabha, ed., *Nation and Narration* (New York: Routledge, 1990); Homi K. Bhabha, *The Location of Culture* (New York: Routledge, 1994).

13. The "black box" here refers to an opaque system of scientific analysis, where only what goes into and what comes out of the system is visible to the outsider. What happens in the black box (i.e., the analysis between input and output) is invisible and unknown to the outsider. In social studies of science, the concept is mostly used with reference to Bruno Latour and his book *Science in Action* (1987).

14. See also Patrick J. Geary, "Genetic History and Migrations in Western Eurasia, 500–1000," in *Empires and Exchanges in Eurasian Late Antiquity*, ed. Nicola Di Cosmo and Michael Maas (Cambridge: Cambridge University Press: 2018), 135–36.

15. See, for example, Paul R. Brewer and Barbara L. Ley, "Media Use and Public Perceptions of DNA Evidence," *Science Communication* 23, no. 1 (2010): 93–117.

Chapter One

1. Carl Zimmer, "The Great Breakup: The First Arrivals to the Americas Split into Two Groups," *New York Times*, May 31, 2018, my emphasis.

2. Most of our physical characteristics are decided by complex interplays between genes in different parts of the genome (which genetic science estimates and calculates as "polygenic scores") and environmental factors of our lives. This means that physical characteristics (phenotypes) cannot be directly "read" from the genome, but must be calculated as degrees of probability. There is also ongoing research on the effects of such "polygenic scores" on our behavior and abilities. For two enthusiastic accounts of the possibilities of the use of DNA for such purposes, see Robert Plomin, *Blueprint: How DNA Makes Us Who We Are* (Cambridge, MA: MIT Press, 2018); and Kathryn Paige Harden, *The Genetic Lottery: Why DNA Matters for Social Equality* (Princeton: Princeton University Press, 2021). For a more nuanced account and excellent summary of the current status and pitfalls of this kind of research, see C. Brandon Ogbunugafor, "DNA, Basketball, and Birthday Luck: A Review of *The Genetic Lottery: Why*

DNA Matters for Social Equality, by Kathryn Paige Harden, 2021," *American Journal of Biological Anthropology* 179 (2022): 501–4.

3. Lynn Hasher, David Goldstein, and Thomas Toppino, "Frequency and the Conference of Referential Validity," *Journal of Verbal Learning and Verbal Behavior* 16, no. 1 (1977): 107–12; Lisa K. Fazio et al., "Knowledge Does Not Protect against Illusory Truth," *Journal of Experimental Psychology* 144, no. 5 (2015): 993–1002.

4. E.g., David Cox, "The Mystery of the Human Genome's Dark Matter," *BBC Future*, April 13, 2023.

5. For a profound yet accessible introduction to the knowledge and history of ideas around DNA and genetics, see Siddhartha Mukherjee, *The Gene: An Intimate History* (London: Bodley Head, 2016).

6. For a discussion of the early relations between anthropology and genetics, see Jonathan Marks, "The Origins of Anthropological Genetics," *Current Anthropology* 53, suppl. 5 (April 2012).

7. For in-depth descriptions and analyses of the role of DNA in popular culture, see José van Dijck, *Imagenation: Popular Images of Genetics* (New York: NYU Press, 1998); and Dorothy Nelkin and M. Susan Lindee, *The DNA Mystique: The Gene as a Cultural Icon* (Ann Arbor: University of Michigan Press, 2004).

8. Van Dijck, *Imagenation*, 2.

9. See also Kim TallBear's similar description of Native American DNA as "a material-semiotic object," in *Native American DNA: Tribal Belonging and the False Promise of Genetic Science* (Minneapolis: University of Minnesota Press, 2013), 17, 54.

10. For more details, see Nalini Gupta, "DNA Extraction and Polymerase Chain Reaction," *Journal of Cytology* 36, no. 2 (2019): 116–17.

11. Gupta, "DNA Extraction and Polymerase Chain Reaction." See also National Human Genome Research Institute, "Polymerase Chain Reaction (PCR) Fact Sheet." https://www.genome.gov/about-genomics/fact-sheets/Polymerase-Chain-Reaction-Fact-Sheet.

12. Russell Higuchi et al., "DNA Sequences from the Quagga, an Extinct Member of the Horse Family," *Nature* 312 (1984): 282–84.

13. Svante Pääbo, "Molecular Cloning of Ancient Egyptian Mummy DNA," *Nature* 314 (1985): 644–45; Svante Pääbo, J. A. Gifford, and Allan C. Wilson, "Mitochondrial DNA Sequences from a 7000-Year-Old Brain," *Nucleic Acids Research* 16, no. 20 (October 1988).

14. Erika Hagelberg, Bryan Sykes, and Robert Hedges, "Ancient Bone DNA Amplified," *Nature* 342, no. 6249 (November 1989).

15. For more extensive reviews of the early developments of the field, see Robert K. Wayne, Jennifer A. Leonard, and Alan Cooper, "Full of Sound and Fury: The Recent History of Ancient DNA," *Annual Review of Ecology and Systematics* 30 (1999): 457–77; Venla Oikkonen, *Population Genetics and Belonging: A Cultural Analysis of Genetic Ancestry* (London: Palgrave Macmillan, 2018), chap. 3;

Elizabeth D. Jones and Elsbeth Bösl, "Ancient Human DNA: A History of Hype (Then and Now)," *Journal of Social Archaeology* 21, no. 2 (2021): 236–55.

16. Matthias Krings et al., "Neanderthal DNA Sequences and the Origin of Modern Humans," *Cell* 90, no. 1 (July 1997).

17. Erika Hagelberg and John Brian Clegg, "Genetic Polymorphisms in Prehistoric Pacific Islanders Determined by Analysis of Ancient Bone DNA," *Proceedings: Biological Sciences* 252, no. 1334 (May 1993); Anne C. Stone and Mark Stoneking, "Ancient DNA from a Pre-Columbian Amerindian Population," *American Journal of Biological Anthropology* 92, no. 4 (December 1993).

18. Olivia Handt et al., "Molecular Genetic Analysis of the Tyrolean Ice Man," *Science* 264, no. 5166 (June 1994).

19. E.g., Alan Cooper and Hendrik N. Poinar, "Ancient DNA: Do It Right or Not At All," *Science* 289, no. 5482 (2000): 1139. See also Jones and Bösl, "Ancient Human DNA: A History of Hype (Then and Now)."

20. Richard E. Green et al., "A Draft Sequence of the Neanderthal Genome," *Science* 328, no. 5979 (May 2010).

21. Wayne, Leonard, and Cooper, "Full of Sound and Fury," 458–59.

22. E.g., Frederika A. Kaestle and K. Ann Horsburgh, "Ancient DNA in Anthropology: Methods, Applications and Ethics," *Yearbook of Physical Anthropology* 45, no. 92 (2002): 106–7 (quotes). See also Keri A. Brown and Mark Pluciennik, "Archaeology and Human Genetics: Lessons for Both," *Antiquity* 75, no. 287 (2001): 101–6.

23. E.g., the ELSI project: https://www.genome.gov/Funded-Programs-Projects/ELSI-Research-Program-ethical-legal-social-implications. See also Anna Källén et al., "Petrous Fever: The Gap between Ideal and Actual Practice in Ancient DNA Research," *Current Anthropology* (2024), and references therein.

24. For a recent example of a bestselling book about DNA that is built on a polarized figure of (the fear of) eugenics and (the hopes of) equality, see Kathryn Paige Harden, *The Genetic Lottery: Why DNA Matters for Social Equality* (Princeton: Princeton University Press, 2021). The balance between hope and fear is a central theme in José van Dijck's book *Imagenation*. See also Mukherjee, *The Gene: An Intimate History*; and Jerome De Groot, *Double Helix History: Genetics and the Past* (New York: Routledge, 2022).

25. Nelkin and Lindee, *The DNA Mystique*, 11.

26. See Mukherjee, *The Gene: An Intimate History*; and K. Ann Horsburgh, "Molecular Anthropology: The Judicial Use of Genetic Data in Archeology." *Journal of Archaeological Science*, 56 (2015): 141–45.

27. Elizabeth D. Jones, *Ancient DNA: The Making of a Celebrity Science* (New Haven: Yale University Press, 2022).

28. Clay Risen, "Bryan Sykes, Who Saw the Ancient Past in Genes, Dies at 73," *New York Times*, January 6, 2021.

29. Bryan Sykes, *The Seven Daughters of Eve* (New York: W. W. Norton, 2001), x.

30. A counterpart from archaeology is Martin Jones, *Unlocking the Past: How*

Archaeologists are Rewriting Human History with Ancient DNA (New York: Arcade Publishing, 2001, 2016).

31. Marks, "The Origins of Anthropological Genetics," S167.

32. The HGP was announced as completed in the year 2000, but twenty years later there were still reports of gaps and missing sections in the map of the "total" human genome. In 2022, there was another announcement of completion. See, for example, Gabrielle Hartley, "The Human Genome Project Pieced Together Only 92% of the DNA—Now Scientists Have Finally Filled In the Remaining 8%," *The Conversation*, March 31, 2022.

33. Text of the White House statements on the Human Genome Project, June 27, 2000. https://archive.nytimes.com/www.nytimes.com/library/national/science/062700sci-genome-text.html.

34. The figure 99.9 percent was later revised, and the mapping of the human genome has proven more complicated than was assumed in the year 2000. For an accessible summary of the complexities, see Emma Kowal and Bastien Llamas, "Race in a Genome: Long Read Sequencing, Ethnicity-Specific Reference Genomes and the Shifting Horizon of Race," *Journal of Anthropological Sciences* 97 (2019): 91–106.

35. E.g., Kowal and Llamas, "Race in a Genome."

36. Hartley, "The Human Genome Project Pieced Together Only 92% of the DNA." See also "Human Genome Project Fact Sheet." https://www.genome.gov/about-genomics/educational-resources/fact-sheets/human-genome-project.

37. Michael Marshall, "Why the Human Genome Was Never Completed," *BBC*, February 13, 2023.

38. See also Charlotte Mulcare and Mélanie Pruvost, "Gained in Translation," in *Critical Perspectives on Ancient DNA*, ed. Daniel Strand, Anna Källén, and Charlotte Mulcare (Cambridge, MA: MIT Press, 2024).

39. See, for example, geneticist Mark Thomas commenting on genetic ancestry tests in "To Claim Someone Has 'Viking Ancestors' Is No Better Than Astrology," *Guardian*, February 25, 2013; and geneticist and anthropologist Jennifer Raff on aDNA and ancestry tests, in "Genetic Astrology: When Ancient DNA Meets Ancestry Testing," *Forbes*, April 9, 2019.

40. Advertisement for genetic ancestry tests from 23andMe. https://www.23andme.com/dna-ancestry.

41. For a more elaborate discussion, see Daniel Strand and Anna Källén, "I Am a Viking! DNA, Popular Culture, and the Construction of Geneticized Identity," *New Genetics and Society* 40, no. 4 (2021): 520–40.

42. The figure is relevant for common analyses of DNA from ancient dry bones, but varies a lot between cases. The amount retrieved in each specific case is indicated by the word "fold" in the scientific paper. There, "fold" represents the percentage of DNA that has been extracted and sequenced, with confidence that the sequence is reproducible. To simplify, you could think of "fold" as roughly the equivalent of percent. A sample that has been analyzed with 1.2

fold is thus based (confidently) on approximately 1.2 percent of the ancient genome.

43. See, for example, Morten Rasmusen et al., "Ancient Human Genome Sequence of an Extinct Palaeo-Eskimo," *Nature* 463, no. 7282 (February 2010); Johannes Krause et al., "The Complete Mitochondrial DNA Genome of an Unknown Hominin from Southern Siberia," *Nature* 464, no. 7290 (2010).

44. Kristian Kristiansen, "Towards a New Paradigm? The Third Science Revolution and Its Possible Consequences in Archaeology," *Current Swedish Archaeology* 22, no. 1 (2014).

45. David Reich, *Who We Are and How We Got Here: Ancient DNA and the New Science of the Human Past* (New York: Oxford University Press, 2018).

46. Ewen Callaway, "The Battle for Common Ground," *Nature* 555 (2018): 573–76.

47. Ann Gibbons, "Revolution in Human Evolution," *Science* 349, no. 6246 (2015).

48. See also Anna Källén, Charlotte Mulcare, Andreas Nyblom, and Daniel Strand, "Introduction: Transcending the aDNA Revolution," *Journal of Social Archaeology* 21, no. 2 (2021): 149–56; Daniel Strand and Anna Källén, "Introduction," in *Critical Perspectives on Ancient DNA*, ed. Daniel Strand, Anna Källén, and Charlotte Mulcare (Cambridge, MA: MIT Press, 2024).

49. Thomas S. Kuhn, *The Structure of Scientific Revolutions* (Chicago: University of Chicago Press, 1996).

50. E.g., Volker Heyd, "Kossinna's Smile," *Antiquity* 91, no. 356 (2017): 348–59; Martin Furholt, "De-contaminating the aDNA–Archaeology Dialogue on Mobility and Migration: Discussing the Culture-historical Legacy," *Current Swedish Archaeology* 17 (2019): 53–68.

51. Here I have taken inspiration from a presentation by geneticist Erika Hagelberg at the Code Narrative History Symposium, May 11, 2021.

52. Reich, *Who We Are*, xxvi.

53. See also Susan E. Hakenbeck, "Genetics, Archaeology and the Far Right: An Unholy Trinity," *World Archaeology* 51, no. 4 (2019): 517–27.

54. Vladimir Putin, "On the Historical Unity of Russians and Ukrainians." http://en.kremlin.ru/events/president/news/66181 (published July 12, 2021; retrieved June 20, 2022).

55. Kristian Kristiansen, "Towards a New Paradigm? The Third Science Revolution and Its Possible Consequences in Archaeology," *Current Swedish Archaeology* 22 (2014): 27.

56. Marianne Sommer, "DNA and Cultures of Remembrance: Anthropological Genetics, Biohistories and Biosocialities," *BioSocieties* 5 (2010): 372.

57. Donna Haraway, "Situated Knowledges: The Science Question in Feminism and the Privilege of Partial Perspective," *Feminist Studies* 14, no 3 (Autumn, 1988): 575–99.

58. Haraway, "Situated Knowledges," 581.

59. E.g., Bruno Latour, *Science in Action: How to Follow Scientists and Engineers through Society* (Cambridge, MA: Harvard University Press, 1987).

60. E.g., Reich, *Who We Are*, xxvi.

61. E.g., Bryan Sykes, *The Seven Daughters of Eve* (New York: W. W. Norton, 2001), x; Gibbons, "Revolution in Human Evolution"; Adam Rutherford, *A Brief History of Everyone Who Ever Lived* (New York: The Experiment, 2017); Reich, *Who We Are*; Kristiansen, "Towards a New Paradigm?" 27.

62. Gideon Lewis-Kraus, "Is Ancient DNA Research Revealing New Truths—or Falling into Old Traps?" *New York Times*, January 17, 2019.

63. Joan Gero, "Honoring Ambiguity/Problematizing Certitude," *Journal of Archaeological Method and Theory* 14 (2007): 311–27.

Chapter Two

1. See https://www.unhcr.org/en-us/news/press/2022/5/628a389e4/unhcr-ukraine-other-conflicts-push-forcibly-displaced-total-100-million.html.

2. See https://www.nobelprize.org/prizes/literature/2021/gurnah/facts/.

3. Bruce G. Trigger, *A History of Archaeological Thought* (Cambridge: Cambridge University Press, 1989), chap. 5 (pp. 148–206).

4. For a theoretical discussion of the arrow as sign in contemporary society, and how it "organises the space into controllable flows," see Gillian Fuller, "The Arrow—Directional Semiotics: Wayfinding in Transit," *Social Semiotics* 12, no. 3 (2002): 231–44.

5. Another example of the use of arrows on maps to explain human and cultural variation in history is reproduced as figure 3 in Marianne Sommer, "Population-genetic Trees, Maps, and Narratives of the Great Human Diasporas," *History of the Human Sciences* 28, no. 5 (2015): 113.

6. For a discussion of Madison Grant's place in the history of anthropology, see Jonathan Marks, "The Origins of Anthropological Genetics," *Current Anthropology* 53, suppl. 5 (2012): S162–63.

7. This trend of dissociation was particularly clear in postwar North American and North European archaeology, where I had my training. The culture-historical perspective has, however, lived on in archaeological communities around the world and has also been continuously promoted by individual archaeologists in Europe and North America.

8. See, for example, Maanasa Raghavan et al., "Genomic Evidence for the Pleistocene and Recent Population History of Native Americans," *Science* 349, no. 6250 (August 21, 2015); Wolfgang Haak et al., "Massive Migration from the Steppe Was a Source for Indo-European Languages in Europe," *Nature* 522 (2015): 217; Alan Cooper and Wolfgang Haak, "European Invasion: DNA Reveals the Origins of Modern Europeans," *The Conversation*, March 22, 2015; Melinda A. Yang, "Ancient DNA Is Revealing the Genetic Landscape of People Who First Settled East Asia," *The Conversation*, September 15, 2020; "Human Migration," in National Geographic Society Resource Library, https://www.nationalgeographic.org/photo/human-migration/; Yichen Liu, Xiaowei Mao,

Johannes Krause, and Qiaomei Fu, "Insights into Human History from the First Decade of Ancient Human Genomics," *Science* 373 (2021): 1479–84. See also the map summarizing "the history of human pigmentation," of Andrea Hanel and Carsten Carlberg, "Skin Colour and Vitamin D: An Update," *Experimental Dermatology* 29 (2020): 868 (fig. 2).

9. Volker Heyd, "Kossinna's Smile," *Antiquity* 91, no. 356 (2017): 348–59; Martin Furholt, "Massive Migrations? The Impact of Recent aDNA Studies on Our View of Third Millennium Europe," *European Journal of Archaeology* 21, no. 2 (2018): 159–91; Martin Furholt, "De-contaminating the aDNA–Archaeology Dialogue on Mobility and Migration: Discussing the Culture-Historical Legacy," *Current Swedish Archaeology* 17 (2019): 53–68; Susan E. Hakenbeck, "Genetics, Archaeology and the Far Right: An Unholy Trinity," *World Archaeology* 51, no. 4 (2019): 517–27.

10. Chao Ning et al., "Ancient Genomes from Northern China Suggest Links between Subsistence Changes and Human Migration," *Nature Communications* 11, no. 2700 (2020).

11. Eske Willerslev and David J. Meltzer, "Peopling of the Americas as Inferred from Ancient Genomics," *Nature* 594 (June 17, 2021), and references therein.

12. In population genetics, a "ghost population" is a hypothetical, but not empirically observed, genetic cluster that is needed to make a map of ancestral relations fit together.

13. These two papers are typical of recent aDNA studies discussing migration, of which there are many similar examples.

14. It should be noted that the terms "whole-genome," "full-genome," and "genome-wide" do not mean that the whole genome has been analyzed, but rather that small sections have been targeted across the full width of the genome. Moreover, as explained in chapter 1, it is often only small fragments of the genome that are preserved and possible to extract and sequence from ancient samples.

15. Chao Ning et al., "Ancient Genomes from Northern China."

16. There are many different genetic terms for mutations—for example, insertions, deletions, microsatellites, and recombination events—but SNP is the term most frequently seen in aDNA studies.

17. But see Kowal and Llamas, "Race in a Genome," for a more detailed account of the complexities of such comparisons.

18. Lisa Loog, "Sometimes Hidden But Always There: The Assumptions Underlying Genetic Inference of Demographic Histories," *Philosophical Transactions of the Royal Society B* 376, no. 1816 (2021).

19. See, for example, John H. Relethford and Deborah A. Bolnick, *Reflections of Our Past: How Human History Is Revealed in Our Genes* (New York: Routledge, 2018).

20. Marc Haber et al., "Ancient DNA and the Rewriting of Human History: Be Sparing with Occam's Razor," *Genome Biology* 17, no. 1 (2016).

21. Catherine J. Frieman and Daniela Hofmann, "Present Pasts in the Archaeology

of Genetics, Identity, and Migration in Europe: A Critical Essay," *World Archaeology* 51, no. 4 (2019): 531–32.

22. Chao Ning et al., "Ancient Genomes from Northern China," 6.

23. Amade M'charek, *The Human Genome Diversity Project: An Ethnography of Scientific Practice* (Cambridge: Cambridge University Press, 2005), 90.

24. See also Kim TallBear, *Native American DNA: Tribal Belonging and the False Promise of Genetic Science* (Minneapolis: University of Minnesota Press, 2013), 7.

25. Benedict Anderson, *Imagined Communities: Reflections on the Origin and Spread of Nationalism*, rev. ed. (London: Verso, 1991), 26 (quote). See also Chris Manias, *Race, Science, and the Nation: Reconstructing the Ancient Past in Britain, France and Germany* (New York: Routledge, 2013).

26. See, for example, Daniel J. Lawson, Lucy van Dorp, and Daniel Falush, "A Tutorial on How Not to Over-interpret STRUCTURE and ADMIXTURE Bar Plots," *Nature Communications* 9, no. 3258 (2018). See also Joanna Brück, "Ancient DNA, Kinship and Relational Identities in Bronze Age Britain," *Antiquity* 95, no. 379 (2021): 228–37.

27. For a similar argument, see Krishna R. Veeramah, "The Importance of Fine-scale Studies for Integrating Paleogenomics and Archaeology," *Current Opinion in Genetics & Development* 53 (2018): 83–89, at 85.

28. E.g., Magnus Fiskesjö, "Ancient DNA and the Politics of Ethnicity in Neo-Nationalist China," in *Critical Perspectives on Ancient DNA*, ed. Daniel Strand, Anna Källén, and Charlotte Mulcare (Cambridge, MA: MIT Press, 2024); Hakenbeck, "Genetics, Archaeology and the Far Right"; Kerstin P. Hofmann, "With *víkingr* into the Identity Trap: When Historiographical Actors get a Life of Their Own," *Medieval Worlds* 4 (2016).

29. See also TallBear, *Native American DNA*, on the construction of Native American genetic identity.

30. Jennifer Raff, *Origins: A Genetic History of the Americas* (New York: Twelve Books, 2022), 176.

31. Frieman and Hofmann, "Present Pasts," 534.

32. Daniel Zadik, "A Handful of Bronze-Age Men Could Have Fathered Two Thirds of Europeans," *The Conversation*, May 21, 2015; George Busby, "Here's How Genetics Helped Crack the History of Human Migration," *The Conversation*, January 13, 2016; Colin Barras, "Story of Most Murderous People of All Time Revealed in Ancient DNA," *New Scientist*, March 27, 2019; Joe Pinkstone, "The Most Violent Group of People Who Ever Lived: Horse-riding Yamnaya Tribe Who Used Their Huge Height and Muscular Build to Brutally Murder and Invade Their Way across Europe More Than 4,000 Years Ago," *Mail Online*, March 29, 2019. See also Frieman and Hofmann, "Present Pasts."

33. Haak et al., "Massive Migration from the Steppe." See also Morten E. Allentoft et al., "Population Genomics of Bronze Age Eurasia," *Nature* 522, no. 7555 (2015): 167–72.

34. Tom Rowsell, "Thank Our Ruthless Barbarian Ancestors for Your Milk," *Why Now*, October 7, 2021.

35. E.g., Allentoft et al., "Population Genomics of Bronze Age Eurasia."

36. Barras, "Story of Most Murderous People."

37. David Reich, *Who We Are and How We Got Here* (New York: Oxford University Press, 2018), 239–40.

38. Abdulrazak Gurnah, *The Last Gift* (London: Bloomsbury, 2011), 242.

39. The RAND Corporation is a US–based global policy think-tank. For the quote, see https://www.rand.org/well-being/portfolios/mass-migration.html.

40. I here follow the most common reference to Denisova as a hominin "species" or "subspecies." However, it should be noted that Svante Pääbo, the geneticist most often connected with the Denisova case, has hesitated to use that term. See, e.g., Matthew Warren, "Mum's a Neanderthal, Dad's a Denisovan: First Discovery of an Ancient-Human Hybrid," *Nature News*, August 22, 2018.

41. E.g., Johannes Krause et al., "The Complete Mitochondrial DNA Genome of an Unknown Hominin from Southern Siberia," *Nature* 464, no. 7290 (2010): 894–97. For a discussion of the scientific and popular production of knowledge about Denisovans, see Mattis Karlsson, *From Fossil to Fact: The Denisova Discovery as Science in Action*, PhD diss., Linköping University, 2022.

42. E.g., Warren, "Mum's a Neanderthal."

43. For an excellent discussion of the Denisova case from a philosophy of science perspective, see Joyce C. Havstad, "Sensational Science, Archaic Hominin Genetics, and Amplified Inductive Risk," *Canadian Journal of Philosophy* 52, no. 3 (2022): 295–320.

44. Bruce Bower, "An Indigenous People in the Philippines Have the Most Denisovan DNA," *Science News*, August 12, 2021.

45. Jamie Shreeve, "The Case of the Missing Ancestor," *National Geographic*, July 2013.

46. Magnus Fiskesjö, "Ancient DNA and the Politics of Ethnicity in Neo-Nationalist China," in *Critical Perspectives on Ancient DNA*, ed. Daniel Strand, Anna Källén, and Charlotte Mulcare (Cambridge, MA: MIT Press, 2024).

47. Michael Greshko, "Ancient DNA Reveals Complex Migrations of the First Americans," *National Geographic*, November 8, 2018. Another great example is *Journey of Man: The Story of the Human Species*, a documentary film directed by Clive Maltby and hosted by Spencer Wells (PBS, 2003).

48. Nick Patterson et al., "Large-scale Migration into Britain during the Middle to Late Bronze Age," *Nature* 601 (2022): 588–94.

49. For a similar argument, see Daniela Hofmann, "What Have Genetics Ever Done for Us?" 460–61.

50. David Melzer and Eske Willerslev, quoted in Ewen Callaway, "Ancient Genomics Is Recasting the Story of the Americas' First Residents," *Nature*, November 8, 2018. See also Pallab Gosh, "DNA Uncovers Mystery Migration to the Americas," *BBC News*, July 2, 2015.

51. Eben Diskin, "Meet the First Man to Walk 14,000 Miles from Argentina to Alaska," *Matador Network*, June 28, 2018.

52. Raff, *Origins*, 207–8. See also Jennifer Raff, "A Genetic Chronicle of the First Peoples in the Americas," *Sapiens*, February 8, 2022.

53. Misia Landau, *Narratives of Human Evolution* (New Haven: Yale University Press, 1993).

54. See, for example, Michel Foucault, *The Order of Things* (New York: Pantheon, 1970), for a thorough discussion of the invention of a new way of thinking about time and origins in modern society in the nineteenth century. It is there that we first find the intense interest in origins in modern science, which we still see today. Hence we can assume that prehistoric people did not have the same understanding of origins as we have today.

55. Fuller, "The Arrow," 239.

56. Mark Thomas, "To Claim Someone Has 'Viking Ancestors' Is No Better Than Astrology," *Guardian*, February 25, 2013.

57. See, for example, Joan Gero, "Honoring Ambiguity/Problematizing Certitude," *Journal of Archaeological Method and Theory* 14 (2007): 311–27; Tim Flohr Sørensen, "In Praise of Vagueness: Uncertainty, Ambiguity and Archaeological Methodology," *Journal of Archaeological Method and Theory* 23 (2016): 741–63.

58. C. Brandon Ogbunugafor, "DNA, Basketball, and Birthday Luck: A Review of *The Genetic Lottery: Why DNA Matters for Social Equality* by Kathryn Paige Harden, 2021," *American Journal of Biological Anthropology* 179 (2022): 502.

59. Hakenbeck, "Genetics, Archaeology and the Far Right."

60. For China, see Jilil Kashgary and Kurban Niyaz, "Chinese Research on Xinjiang Mummies Seen As Promoting Revisionist History," *Radio Free Asia*, June 11, 2022, https://www.rfa.org/english/news/uyghur/melting-pot -06032022104308.html. For Greece, see Yannis Hamilakis, "Who Are You Calling Mycenaean?" *London Review of Books Blog*, August 10, 2017. For Israel, see Megan Gannon, "When Ancient DNA Gets Politicized," *Smithsonian Magazine*, July 12, 2019. For the United States, see Lizzie Wade, "Ancient DNA Confirms Native Americans' Deep Roots in North and South America," *Science*, November 8, 2018.

Chapter Three

1. The well-known US senator from Vermont. See https://www.igenea.com/ en/famous-people. iGENEA is just one example of many companies offering genetic ancestry tests with similar sales rhetorics. Other examples are AncestryDNA, 23andMe, MyHeritage, and FamilyTreeDNA.

2. Venla Oikkonen, "Mitochondrial Eve and the Affective Politics of Human Ancestry," *Signs: Journal of Women in Culture and Society* 40, no. 3 (2015).

3. Mark Thomas, "To Claim Someone Has 'Viking Ancestors' Is No Better Than Astrology," *Guardian*, February 25, 2013.

4. See, for example, Kostas Kampourakis, *Dismantling the Myth of Genetic Ethnicities* (New York: Oxford University Press, 2023); Alondra Nelson, *The Social Life of DNA: Race, Reparations, and Reconciliation after the Genome* (Boston: Beacon Press, 2016); Jackie Hogan, *Roots Quest: Inside America's Genealogy Boom* (Lanham, MD: Rowman & Littlefield, 2019); Catherine Nash, *Genetic Geographies: The Trouble with Ancestry* (Minneapolis: University of Minnesota Press, 2015); Deborah A. Bolnick et al., "The Science and Business of Genetic Ancestry Testing," *Science* 318, no. 19 (2007): 399–400; Mark Jobling, R. Rasteiro, and J. H. Wetton, "In the Blood: The Myth and Reality of Genetic Markers of Identity," *Ethnic and Racial Studies* 39, no. 2 (2016): 142–61; "Debunking Genetic Astrology," https://www.ucl.ac.uk/biosciences/molecular-and-cultural-evolution -lab/debunking-genetic-astrology; Jennifer Raff, "Genetic Astrology: When Ancient DNA Meets Ancestry Testing," *Forbes*, April 9, 2019.

5. See, for example, Henry Louis Gates Jr., *Finding Oprah's Roots: Finding Your Own* (New York: Skyhorse Publishing, 2007); and the story of Las Abuelas de Plaza de Mayo, in Nelson, *The Social Life of DNA*, 28–32.

6. Around 100 million kits for genetic ancestry tests had been sold by the end of 2021, and the market was expected to grow in the coming years, according to a 2021 report by Research and Markets. https://www.researchandmarkets.com/ reports/5351043.

7. Chip Colwell, "Is DNA a Dangerous Heritage?" in *Polarized Pasts: Heritage and Belonging in Times of Political Polarization*, ed. Elisabeth Niklasson (New York: Berghahn Books, 2023), 374–418. See also "Elizabeth Warren: DNA Test Finds 'Strong Evidence' of Native American Blood," *BBC News*, October 15, 2018.

8. Matt Viser, "Donald Trump's Drive to Surpass His Father's Success," *Boston Globe*, July 16, 2016.

9. Jenny Reardon, "The Democratic, Anti-racist Genome? Technoscience at the Limits of Liberalism," *Science As Culture* 21, no. 1 (2012): 38–40.

10. Oprah Winfrey, quoted at https://www.oprah.com/oprahshow/dr-henry-louis -gates-jr-start-your-ancestry-search/all.

11. David Brown, "Revealed: The Indian Ancestry of William," *Times* (London), June 14, 2013. See also Daniel Zadik, "Will-I-am Indian, But Does It Matter?" *The Conversation*, June 14, 2013.

12. Sahm Venter, "DNA Test May Reveal You're Related to Madiba," *IOL*, March 7, 2006.

13. For an empirically grounded analysis, see Daniel Strand and Anna Källén, "I Am a Viking! DNA, Popular Culture, and the Construction of Geneticized Identity," *New Genetics and Society* 40, no. 4 (2021): 520–40.

14. This does not mean that race-thinking or racism does not exist in Sweden, but the word "race" is not used in the same way and with the same connotations as in other contexts. For a more elaborate argument, see Anna Källén, "The Sigtuna Debacle," in *Polarized Pasts: Heritage and Belonging in Times of Political Polarization*, ed. Elisabeth Niklasson (New York: Berghahn Books, 2023).

15. Magnus Fiskesjö, "Ancient DNA and the Politics of Ethnicity in Neo-Nationalist China," in *Critical Perspectives on Ancient DNA*, ed. Daniel Strand, Anna Källén, and Charlotte Mulcare (Cambridge, MA: MIT Press, 2024).

16. For a similar argument on genetics and identity in the Middle East, see Ian McGonigle, *Genomic Citizenship: The Molecularization of Identity in the Contemporary Middle East* (Cambridge, MA: MIT Press, 2021).

17. Kim TallBear, *Native American DNA: Tribal Belonging and the False Promise of Genetic Science* (Minneapolis: University of Minnesota Press, 2013), 1–2.

18. TallBear, *Native American DNA*, 4.

19. TallBear, *Native American DNA*, 47–63. For a similar example from the context of the Sámi in Scandinavia, see Mikkel Berg-Nordlie, "'Sámi in the Heart': Kinship, Culture, and Community as Foundations for Indigenous Sámi Identity in Norway," *Ethnopolitics* 21, no. 4 (2022): 450–72.

20. E.g., Julie M. Schablitsky et al., "Ancient DNA Analysis of a Nineteenth-century Tobacco Pipe from a Maryland Slave Quarter," *Journal of Archaeological Science* 105 (2019): 11–18; Raquel E. Fleskes et al., "Community-engaged Ancient DNA Project Reveals Diverse Origins of 18th-century African Descendants in Charleston, South Carolina," *PNAS* 120, no. 3 (2023); see also examples in Nelson, *The Social Life of DNA*.

21. Songül Alpaslan-Roodenberg et al., "Ethics of DNA Research on Human Remains: Five Globally Applicable Guidelines," *Nature Communication* (2021).

22. For broad accounts of the HGDP, see Jenny Reardon, *Race to the Finish: Identity and Governance in an Age of Genomics* (Princeton: Princeton University Press, 2004); Amade M'charek, *The Human Genome Diversity Project: An Ethnography of Scientific Practice* (Cambridge: Cambridge University Press, 2005); Jonathan Marks, "'We're Going to Tell These People Who They Really Are': Science and Relatedness," in *Relative Values: Reconfiguring Kinship Studies*, ed. S. Franklin and S. McKinnon (Durham, NC: Duke University Press, 2002), 355–83.

23. L. Luca Cavalli-Sforza et al., "Call for a Worldwide Survey of Human Genetic Diversity: A Vanishing Opportunity for the Human Genome Project," *Genomics* 11 (Summer 1991): 490–91.

24. L. Luca Cavalli-Sforza, Paolo Menozzi, and Alberto Piazzi, *The History and Geography of Human Genes* (Princeton: Princeton University Press, 1994). See also Jenny Reardon, "The Human Genome Diversity Project: What Went Wrong?" in *The Postcolonial Science and Technology Studies Reader*, ed. Sandra Harding (Durham, NC: Duke University Press, 2011), 321–42.

25. But see also the landmark "Mitochondrial Eve" study of human evolution: Rebecca L. Cann, Mark Stoneking, and Allan C. Wilson, "Mitochondrial DNA and Human Evolution," *Nature* 325 (1987): 31–36.

26. Some of these ethnonyms, including "Lapp," are now considered inappropriate and offensive.

27. Marianne Sommer, *History Within: The Science, Culture, and Politics of Bones,*

Organisms, and Molecules (Chicago: University of Chicago Press, 2016); Marks, "'We're Going to Tell These People Who They Really Are.'"

28. For accounts of criticism of the HGDP, see, for example, Reardon, "The Human Genome Diversity Project: What Went Wrong?"; Marks, "'We're Going to Tell These People Who They Really Are'"; Donna J. Haraway, *Modest_Witness@ Second_Millennium.FemaleMan©_meets_OncoMouse™: Feminism and Technoscience* (New York: Routledge, 1997).

29. Leslie Roberts, "A Genetic Survey of Vanishing Peoples," *Science* 252, no. 5013 (June 21, 1991): 1614.

30. Patricia Kahn, "Genetic Diversity Project Tries Again," *Science* 266, no. 5186 (November 4, 1994): 721; Charles J. Hanley, "Indigenous Peoples Resist Worldwide Gene Study," *Los Angeles Times*, July 7, 1996.

31. Kahn, "Genetic Diversity Project Tries Again," 720.

32. Roberts, "A Genetic Survey of Vanishing Peoples," 1614.

33. Marks, "'We're Going to Tell These People Who They Really Are,'" 369.

34. Kahn, "Genetic Diversity Project Tries Again," 721.

35. L. Luca Cavalli-Sforza, "The Human Genome Diversity Project: Past, Present and Future," *Nature Reviews Genetics* 6 (April 2005): 333.

36. Reardon, *Race to the Finish*; Reardon, "The Human Genome Diversity Project: What Went Wrong?"

37. Marks, "'We're Going to Tell These People Who They Really Are,'" 373–74.

38. Cavalli-Sforza et al., "Call for a Worldwide Survey of Human Genetic Diversity."

39. For more thorough descriptions and analyses of the Genographic Project, see Sommer, *History Within*, 314–30, 345–54; Catherine Nash, "Genetics, Race, and Relatedness: Human Mobility and Human Diversity in the Genographic Project," *Annals of the Association of American Geographers* 102, no. 3 (May 2012): 667–84; TallBear, *Native American DNA*, chap. 4; Nash, *Genetic Geographies*, chap. 2.

40. Genographic Project, cited in Nash, "Genetics, Race, and Relatedness," 667.

41. See https://www.nationalgeographic.com/pages/article/genographic.

42. There was indeed some criticism of the Genographic Project similar to criticism of the HGDP, but it did not have the same resounding effects. See, e.g., Ellen L. Lutz, "Cultural Survival Calls for Genographics Moratorium," *Cultural Survival Quarterly Magazine* 30, no. 9 (September 2006); Jenny Reardon and Kim TallBear, "'Your DNA Is Our History': Genomics, Anthropology, and the Construction of Whiteness as Property," *Current Anthropology* 53, suppl. 5 (2012).

43. "Decoding Implications of the Genographic Project for Archaeology and Cultural Heritage" (transcript of a panel discussion held at the Chacmool Conference "Decolonizing Archaeology," University of Calgary, Alberta, Canada, November 2006, convened by George Nicholas and Julie Hollowell), *International Journal of Cultural Property* 16 (2009): 141–81.

44. Harry, in "Decoding Implications," 147.

45. *Journey of Man: The Story of the Human Species*, documentary film directed by Clive Maltby and hosted by Spencer Wells (PBS, 2003).

46. The Genographic Project® Geno 2.0 Next Generation Helix Product Privacy Policy. https://www.nationalgeographic.com/pages/article/genographic.

47. Reich, *Who We Are and How We Got Here* (New York: Oxford University Press, 2018), xx–xxi.

48. E.g., Swapan Mallick et al., "The Simons Genome Diversity Project: 300 Genomes from 142 Diverse Populations," *Nature* 538 (October 13, 2016): 201–6.

49. Reardon and TallBear, "'Your DNA Is Our History,'" S243–44; Emma Kowal et al., "Community Partnerships Are Fundamental to Ethical Ancient DNA Research," *HGG Advances* 4, no. 2 (2023): 100161; Jessica Bardill et al., "Advancing the Ethics of Paleogenomics," *Science* 360, no. 6387 (2018): 384–85; Jennifer K. Wagner et al., "Fostering Responsible Research on Ancient DNA," *American Journal of Human Genetics* 107, no. 2 (2020): 183–95. See also Amanda Height, "Ancient DNA Boom Underlines a Need for Ethical Frameworks," *Scientist*, January 27, 2022; Nicky Phillips, "Indigenous Groups Look to Ancient DNA to Bring Their Ancestors Home," *Nature* 568 (2019): 294–97; Lizzie Wade, "Ancient DNA Confirms Native Americans' Deep Roots in North and South America," *Science*, November 8, 2018; Chip Colwell, "Rights of the Dead and the Living Clash When Scientists Extract DNA from Human Remains," *The Conversation*, April 6, 2018; Jennifer Raff, *Origins: A Genetic History of the Americas* (New York: Twelve Books, 2022).

50. Reardon, *Race to the Finish*; Reardon, "The Democratic, Anti-racist Genome?," 38–40.

51. Yulia Egorova, "DNA Evidence? The Impact of Genetic Research on Historical Debates," *BioSocieties* 5, no. 3 (2010): 348–65.

52. See Anna Källén et al., "Petrous Fever: The Gap between Ideal and Actual Practice in Ancient DNA Research," *Current Anthropology* (2024), and references therein.

53. For a more extensive argument, see Källén et al., "Petrous Fever." See also María C. Ávila Arcos, "Wielding New Genomic Tools Wisely: Troubling Traces of Biocolonialism Undermine an Otherwise Eloquent Synthesis of Ancient Genome Research," *Science* 360, no. 6386 (April 20, 2018).

54. For a more detailed description of the meaning of "admixture" (as a process and a state), see Marianne Sommer and Ruth Amstutz, "Diagrams of Human Genetic Kinship and Diversity," in *Critical Perspectives on Ancient DNA*, ed. Daniel Strand, Anna Källén, and Charlotte Mulcare (Cambridge, MA: MIT Press, 2024).

55. TallBear, *Native American DNA*, 5.

56. Daniel Strand: "0.01%: Genetics, Race and the Methodology of Differentiation," *Eurozine*, January 1, 2021.

57. The simple tree shape has been revised in later works (e.g. Reich, *Who We Are and How We Got Here*), but remains as a resilient image in genetic science and

popular communication. See, for example, the recent scientific and popular articles by Yan Wong, a TV personality and an evolutionary geneticist at Oxford University's Big Data Institute: Yan Wong et al., "A Unified Genealogy of Modern and Ancient Genomes," *Science* 375, no. 6583 (February 25, 2022); and Yan Wong and Anthony Wilder Wohns, "We're Analysing DNA from Ancient and Modern Humans to Create a 'Family Tree of Everyone,'" *The Conversation*, February 28, 2022.

58. E.g., Sommer, *History Within*; Petter Hellström, "Trees of Knowledge: Science and the Shape of Genealogy," PhD diss., Uppsala University, 2019; Jonathan Marks, "The Origins of Anthropological Genetics," *Current Anthropology* 53, suppl. 5 (April 2012).

59. See Sommer and Amstutz, "Diagrams of Human Genetic Kinship and Diversity."

60. This was pointed out as a problem already by Cavalli-Sforza in the 1970s, and population geneticists have since called for the development of alternative illustration models, based on cluster or network shapes (see Sommer and Amstutz, "Diagrams of Human Genetic Kinship and Diversity"), but the tree model remains in circulation as a common illustration of the genetic relations of humanity.

61. Sommer and Amstutz, "Diagrams of Human Genetic Kinship and Diversity." See also Marks, "'We're Going to Tell These People Who They Really Are.'"

62. Along with the increasingly common cluster illustrations.

63. Marianne Sommer, "Population-genetic Trees, Maps, and Narratives of the Great Human Diasporas," *History of the Human Sciences* 28, no. 5 (2015): 108–45.

64. See discussion above, as well as in Berg-Nordlie, "'Sámi in the Heart'"; and TallBear, *Native American DNA*.

65. E.g., Nicky Phillips, "Indigenous Groups Look to Ancient DNA to Bring Their Ancestors Home," *Nature* 568 (2019): 294–97; Morten Rasmussen et al., "The Ancestry and Affiliations of Kennewick Man," *Nature* 523 (July 2015): 455–58; Joanne L. Wright et al., "Ancient Nuclear Genomes Enable Repatriation of Indigenous Human Remains," *Science Advances* 4 (2018). See also Emma Kowal, *Haunting Biology: Science and Indigeneity in Australia* (Durham, NC: Duke University Press, 2023), and Sarah Wild, "Does DNA Simplify or Complicate Repatriation Claims?" *Sapiens*, December 8, 2021.

66. Ernie LaPointe, *Sitting Bull: His Life and Legacy* (Layton, UT: Gibbs Smith, 2009); *Sitting Bull's Voice*, documentary film produced by Ernie LaPointe, Bill Matson, and Tim Prokop (2014).

67. John Travis, "In Search of Sitting Bull," *Science* 330 (October 8, 2010): 172–73; John Travis, "Honoring His Ancestor by Studying His DNA," *Science* 330, no. 6001 (2010); *Lost Bones: In Search of Sitting Bull's Grave*, documentary film produced by Kyle Bornais and Dave Gaudet, featuring Ernie LaPointe (Farpoint Films, 2009).

68. Travis, "In Search of Sitting Bull," 173.

69. Travis, "In Search of Sitting Bull" and "Honoring His Ancestor by Studying His DNA."

70. Jaqueline Garget, "Living Descendant of Sitting Bull Confirmed by Analysis of DNA from the Legendary Leader's Hair," University of Cambridge News Release, October 27, 2021. https://www.cam.ac.uk/stories/sitting-bull-descendant-confirmed. For news reports, see, for example, Katie Hunt, "Sitting Bull's Great-grandson Identified Using New DNA Technique," *CNN*, October 28, 2021.

71. Ida Moltke et al., "Identifying a Living Great-grandson of the Lakota Sioux Leader Tatanka Iyotake (Sitting Bull)," *Science Advances* 7, no. 44 (October 27, 2021).

72. Mikkel Berg-Nordlie discusses in similar terms the use of DNA as metaphor in the context of Sámi identity, in Berg-Nordlie, "'Sámi in the Heart.'"

73. Sénat Coutumier Nouvelle-Calédonie, "Délibération relative aux recueils d'acide desoxyribonucleïque (ADN) et ribonucleïque (ARN) sur des sujets autochtones (Kanak) en Nouvelle-Calédonie, à des fins non médicales," May 2017. See also Christophe Sand, commentary in forum "Ancient DNA and Its Contribution to Understanding the Human History of the Pacific Islands," *Archaeology in Oceania* 53 (2018): 214–15.

74. Phillips, "Indigenous Groups Look to Ancient DNA"; Rasmussen et al., "The Ancestry and Affiliations of Kennewick Man"; Wright et al., "Ancient Nuclear Genomes Enable Repatriation"; Wild, "Does DNA Simplify or Complicate"; Height, "Ancient DNA Boom."

75. E.g., Amanda Daniela Cortez et al., "An Ethical Crisis in Ancient DNA Research: Insights from the Chaco Canyon Controversy As a Case Study," *Journal of Social Archaeology* 21, no. 2 (2021): 157–78.

76. E.g., Stewart B. Koyiyumptewa et al., "Twisting Strings: Hopi Ancestors and Ancient DNA," in *Critical Perspectives on Ancient DNA*, ed. Daniel Strand, Anna Källén, and Charlotte Mulcare (Cambridge, MA: MIT Press, 2024); Berg-Nordlie, "'Sámi in the Heart.'" See also Marie Annette Jaimes Guerrero, "Biocolonialism and Isolates of Historic Interest," in *Indigenous Intellectual Property Rights: Legal Obstacles and Innovative Solutions*, ed. Mary Riley (Walnut Creek, CA: Altamira Press, 2004), 251–77; TallBear, *Native American DNA*; Reardon and TallBear, "'Your DNA Is Our History,'" S233–34.

77. E.g., Guerrero, "Biocolonialism."

78. E.g., Raff, *Origins*; Michael Greshko, "Ancient DNA Reveals Complex Migrations of the First Americans," *National Geographic*, November 8, 2018; Carl Zimmer, "The Great Breakup: The First Arrivals to the Americas Split into Two Groups," *New York Times*, May 31, 2018.

79. E.g., Jessica Bardill et al., "Advancing the Ethics of Paleogenomics," *Science* 360, no. 6387 (2018): 384–85; Jennifer K. Wagner et al., "Fostering Responsible Research on Ancient DNA," *American Journal of Human Genetics* 107, no. 2 (2020): 183–95.

80. E.g., Rasmussen et al., "The Ancestry and Affiliations of Kennewick Man"; Raff, *Origins*.

81. For a thorough discussion of the historical role of genetic and other science in the creation and maintenance of hegemonic structures of inequality and the marginalization of Indigenous people in Australia, see Kowal, *Haunting Biology*.

82. E.g., Koyiyumptewa et al., "Twisting Strings: Hopi Ancestors and Ancient DNA."

83. *Lost Bones*, documentary film.

84. Moltke et al., "Identifying a Living Great-grandson," 3.

85. See the discussion of the term "admixture" in Nicholas V. Passalacqua and Marin A. Pilloud, *Ethics and Professionalism in Forensic Anthropology* (Amsterdam: Elsevier, 2018), 91–92. See also Strand, "0.01%."

86. Margaret Sleeboom-Faulkner, "How to Define a Population: Cultural Politics and Population Genetics in the People's Republic of China and the Republic of China," *BioSocieties* 1 (2006): 349–419. See also Fiskesjö, "Ancient DNA and the Politics of Ethnicity."

87. Vikas Kumar et al., "Bronze and Iron Age Population Movements Underlie Xinjiang Population History," *Science* 376 (2022): 62–69.

88. Amnesty International, *"Like We Were Enemies in a War": China's Mass Internment, Torture and Persecution of Muslims in Xinjiang*, https://xinjiang.amnesty.org; United Nations, *OHCHR Assessment of Human Rights Concerns in the Xinjiang Uyghur Autonomous Region, People's Republic of China*, UN Country Reports, August 31, 2022.

89. Jilil Kashgary and Kurban Niyaz, "Chinese Research on Xinjiang Mummies Seen As Promoting Revisionist History," *Radio Free Asia*, June 11, 2022.

90. See also Källén, "The Sigtuna Debacle."

91. Kumar et al., "Bronze and Iron Age," 3.

92. Chao Ning et al., "Ancient Genomes from Northern China Suggest Links between Subsistence Changes and Human Migration," *Nature Communications* 11, no. 2700 (2020): 5.

93. E.g., Mark Munsterhjelm, *Forensic Colonialism: Genetics and the Capture of Indigenous People* (Montreal: McGill-Queen's University Press, 2023); Kashgary and Niyaz, "Chinese Research on Xinjiang Mummies"; Fiskesjö, "Ancient DNA and the Politics of Ethnicity."

94. Fiskesjö, "Ancient DNA and the Politics of Ethnicity."

95. Reardon, *Race to the Finish*; Reardon, "The Democratic, Anti-racist Genome?," 38–40.

96. E.g., Koyiyumptewa et al., "Twisting Strings: Hopi Ancestors and Ancient DNA"; Berg-Nordlie, " 'Sámi in the Heart.' "

97. For a similar argument, see Kerstin P. Hofmann, "With *Víkingr* into the Identity Trap: When Historiographical Actors Get a Life of Their Own," *Medieval Worlds* 4 (2016): 92.

Chapter Four

1. *The First Brit: Secrets of the 10,000-Year-Old Man*, Channel 4 documentary, first aired on February 18, 2018. Depicted scene around 38.30–39.30 min. See also https://www.nhm.ac.uk/press-office/press-releases/the-first-brit--secrets-of-the-10-000-year-old-man.html. The consulted expert was Dr. Susan Walsh. See, e.g., Susan Walsh et al., "Global Skin Colour Prediction from DNA," *Human Genetics* 136 (2017): 847–63; Lakshmi Chaitanya et al., "The HIrisPlex-S System for Eye, Hair and Skin Colour Prediction from DNA: Introduction and Forensic Developmental Validation," *Forensic Science International: Genetics* 35 (2018): 123–35.

2. The information about the excavations and subsequent treatment of the skeletal remains is taken from Henry N. Davies, "The Discovery of Human Remains under the Stalagmite Floor of Gough's Cavern, Cheddar," *Quarterly Journal of the Geological Society* (August 1904): 335–48; Roger M. Jacobi, "The History and Literature of Pleistocene Discoveries at Gough's Cave, Cheddar, Somerset," *Proceedings of the University of Bristol Spelaeological Society* 17, no. 2 (1985): 102–15; and the website (with personal testimonies and original photographs) http://www.nonesuchexpeditions.com/nonesuch-features/cheddar_man/Cheddar_Man.htm.

3. See original photographs and comment in Davies, "The Discovery," pl. XXIX and p. 347.

4. The postcard is reproduced in Jacobi, "The History," pl. 1.

5. For the context of rivalry between France and Britain, see Murray Goulden, "Bringing Bones to Life: How Science Made Piltdown Man Human," *Science as Culture* 16, no. 4 (2007): 333–57; Marianne Sommer, "Mirror, Mirror on the Wall: Neanderthal as Image and 'Distortion' in Early 20th-Century French Science and Press," *Social Studies of Science* 36, no. 2 (2006): 207–40; and Venla Oikkonen, *Population Genetics and Belonging: A Cultural Analysis of Genetic Ancestry* (London: Palgrave Macmillan, 2018), 102.

6. See http://www.nonesuchexpeditions.com/nonesuch-features/cheddar_man/Cheddar_Man.htm; Oikkonen, *Population Genetics and Belonging*, 102; Jacobi, "The History," 105.

7. See https://www.nhm.ac.uk/discover/cheddar-man-mesolithic-britain-blue-eyed-boy.html; https://www.nhm.ac.uk/press-office/press-releases/the-first-brit--secrets-of-the-10-000-year-old-man.html.

8. There is to my knowledge no scientific paper backing this study, but some of the procedure is described in Bryan Sykes, *The Seven Daughters of Eve* (London: Corgi Books, 2002), chap. 12. The context is also described and analyzed in Oikkonen, *Population Genetics and Belonging*, 102–19.

9. Sykes, *The Seven Daughters of Eve*, 220–21.

10. E.g., Hans-Jürgen Bandelt et al., "The Brave New Era of Human Genetic Testing," *BioEssays* 30, nos. 11–12 (2008): 1246–51; Kerstin P. Hofmann, "With

víkingr into the Identity Trap: When Historiographical Actors Get a Life of Their Own," *Medieval Worlds* 4 (2016): 98–99.

11. E.g., Dana Kristjansson et al., "Evolution and Dispersal of Mitochondrial DNA Haplogroup U5 in Northern Europe: Insights from an Unsupervised Learning Approach to Phylogeography," *BMC Genomics* 23 (2022): art. no. 354.

12. For a critical rendering of the analysis, see Bandelt et al., "The Brave New Era," 1247.

13. For more examples and an excellent analysis of affect, belonging, and national identity in the media coverage of Targett's relation to Cheddar Man, see Oikkonen, *Population Genetics and Belonging*, 105–8, and references therein. The relation has also been noted in the *Guinness Book of World Records*, where Targett is said to have "been linked through some 300 generations and been shown to be a direct descendant [*sic*], on his mother's side, of Cheddar Man." https://www.guinnessworldrecords.com/world-records/67457-farthest-traced -descendant-by-dna.

14. A photograph of the Manchester University team's bust model of Cheddar Man can be seen at https://www.nhm.ac.uk/discover/news/2018/january/ documentary-to-reveal-surprising-face-of-cheddar-man.html.

15. Catherine Targett was quoted in the *Bristol Evening Post*, March 6, 1998, and is cited in Oikkonen, *Population Genetics and Belonging*, 110.

16. For a critical perspective on the use of prehistoric archaeology, including Cheddar Man, in the context of Brexit, see Kenneth Brophy, "The Brexit Hypothesis and Prehistory," *Antiquity* 92, no. 366 (2018): 1650–58. See also Chiara Bonacchi et al., "The Heritage of Brexit: Roles of the Past in the Construction of Political Identities through Social Media," *Journal of Social Archaeology* 18 (2018): 174–92.

17. *The First Brit*, documentary. Speaker voice: Jim Carter. Depicted scene around 0.00–0.58 min. Emphasis added to reflect emphasis in the reading.

18. E.g., Ron Pinhasi et al., "Optimal Ancient DNA Yields from the Inner Ear Part of the Human Petrous Bone," *PLoS ONE* 10 (2015).

19. *The First Brit*, 10.09–11.06 min.

20. Sykes, *The Seven Daughters of Eve*, 223–24.

21. *The First Brit*, 12.00–12.18 min.

22. Selena Brace et al., "Ancient Genomes Indicate Population Replacement in Early Neolithic Britain," *Nature Ecology & Evolution* 3 (May 2019): 765–71.

23. For a short summary of the complicated relations between genotype and phenotype, and of the importance of the *meanings and values* we ascribe to phenotypes, see C. Brandon Ogbunugafor, "DNA, Basketball, and Birthday Luck: A Review of *The Genetic Lottery: Why DNA Matters for Social Equality* by Kathryn Paige Harden, 2021," *American Journal of Biological Anthropology* 179 (2022): 501–4. See also John Hawks's blog post, "What Color Were Neandertals," at https://johnhawks.net/weblog/what-color-were-neandertals/.

24. Susan Walsh and Manfred Kayser, in supplementary materials for Brace et al.,

"Ancient Genomes," 18–19; Chaitanya et al., "The HIrisPlex-S System." See also Nina G. Jablonski, "The Evolution of Human Skin Pigmentation Involved the Interactions of Genetic, Environmental, and Cultural Variables," *Pigment Cell & Melanoma Research* 34, no. 4 (2021): 707–29.

25. Kit Buchan, "Meet the Ancestors . . . The Two Brothers Creating Lifelike Figures of Early Man," *Guardian*, May 5, 2018.

26. Buchan, "Meet the Ancestors . . ."

27. *The First Brit*, 43.30–43.50 min.

28. Walsh and Kayser, in supplementary materials for Brace et al., "Ancient Genomes," 18–19.

29. Walsh and Kayser, in supplementary materials for Brace et al., "Ancient Genomes," 18–19, emphasis added.

30. Kerry Lotzof, "Cheddar Man: Mesolithic Britain's Blue-eyed Boy," on Natural History Museum website, https://www.nhm.ac.uk/discover/cheddar-man -mesolithic-britain-blue-eyed-boy.html.

31. *The First Brit*, 44.00–44.05 min. This can be compared with a similar coming-to-life description of Cheddar Man's skull in Sykes, *The Seven Daughters of Eve*, 220–21.

32. Steven Morris, "'He's One of Us': Modern Neighbours Welcome Cheddar Man," *Guardian*, February 9, 2018.

33. John Naish, "Meet My Ancestor Cheddar Man," *Daily Mail*, February 7, 2018.

34. See a collage of Neandertal faces in Hawks, "What Color Were Neandertals."

35. Michael Balter, "Mystery Solved: 8500-year-old Kennewick Man Is a Native American after All," *Science*, June 18, 2015; see also the original article by Morten Rasmussen et al., "The Ancestry and Affiliations of Kennewick Man," *Nature* 523 (2015): 455–58. For context, see also Jennifer Raff, *Origins: A Genetic History of the Americas* (New York: Twelve Books, 2022), chap. 9.

36. Andreas Keller et al., "New Insights into the Tyrolean Iceman's Origin and Phenotype As Inferred by Whole-genome Sequencing," *Nature Communications* 3, no. 698 (February 28, 2012); Ewen Callaway, "Iceman's DNA Reveals Health Risks and Relations," *Nature News*, February 28, 2012. See also John Robb, "Towards a Critical Ötziography: Inventing Prehistoric Bodies," in *Social Bodies*, ed. Helen Lambert and Maryon McDonald (New York: Berghahn Books, 2009), 100–128.

37. Pere Gelabert et al., "Northeastern Asian and Jomon-related Genetic Structure in the Three Kingdoms Period of Gimhae, Korea," *Current Biology* 32 (August 8, 2022): 3232–44; https://pub.parabon.com/Parabon-Snapshot -Scientific-Poster—ISHI-2021—DNA-Phenotyping-on-Ancient-DNA-from -Egyptian-Mummies.pdf.

38. Brian Handwerk, "Scientists Recreate the Face of a Denisovan Using DNA," *Smithsonian Magazine*, September 19, 2019; see also the original article David Gokhman et al., "Reconstructing Denisovan Anatomy Using DNA Methylation Maps," *Cell*, 179 (September 19, 2019): 180–92.

39. Caroline Wilkinson, *Forensic Facial Reconstruction* (Cambridge: Cambridge University Press, 2004), 39–40. See also Laura Buti et al., "Facial Reconstruction of Famous Historical Figures: Between Science and Art," in *Studies in Forensic Biohistory*, ed. Christopher M. Stojanowski and William N. Duncan (Cambridge: Cambridge University Press, 2017), 191–212.

40. Robb, "Towards a Critical Ötziography," 105.

41. Gokhman et al., "Reconstructing Denisovan Anatomy."

42. Gokhman et al., "Reconstructing Denisovan Anatomy," 187.

43. Sommer, "Mirror, Mirror," 225–30; Stephanie Moser, *Ancestral Images: The Iconography of Human Origins* (Ithaca, NY: Cornell University Press, 1998).

44. Michael Price, "Ancient DNA Puts a Face on the Mysterious Denisovans, Extinct Cousins of Neanderthals," *Science*, September 19, 2019; Ewen Callaway, "Denisovan Portrait Drawn from DNA," *Nature*, September 26, 2019: 475–76.

45. See, for example, Julie D. White et al., "Insights into the Genetic Architecture of the Human Face," *Nature Genetics* 53 (January 2021): 45–53.

46. E.g., Amade M'charek and Katharina Schramm, "Encountering the Face—Unraveling Race," *American Anthropologist* 122 (2020): 321–26; Abigail Nieves Delgado, "The Problematic Use of Race in Facial Reconstruction," *Science as Culture* 29 (2020): 568–93; Carrie Arnold, "Crimefighting with Family Trees," *Nature* 585 (September 10, 2020).

47. Roos Hopman, "The Face As Folded Object: Race and the Problems with 'Progress' in Forensic DNA Phenotyping," *Social Studies of Science* (2021): 3.

48. UNESCO Declaration on Race and Racial Prejudice (1979), 23. https://unesdoc .unesco.org/ark:/48223/pf0000039429. See also Reardon, *Race to the Finish*, chap. 2, for a thorough description and analysis of the complicated and far from unified processes that led to the eventual formulation of the statement.

49. E.g., David Reich, *Who We Are and How We Got Here* (New York: Oxford University Press, 2018).

50. Adam Rutherford, "A Cautionary History of Eugenics," *Science* 373, no. 6562 (2021): 1419; Kathryn Paige Harden, *The Genetic Lottery: Why DNA Matters for Social Equality* (Princeton: Princeton University Press, 2021).

51. For examples of technologies of visualization in earlier race science that can now be seen "folded into" current practices of aDNA research, see Veronika Lipphardt, "Traditions and Innovations: Visualizations of Human Variation, c. 1900–38," *History of the Human Sciences* 28, no. 5 (2015): 49–79. See also Ann Morning speaking of a "genetic black-boxing of race," in "And You Thought We Had Moved beyond All That: Biological Race Returns to the Social Sciences," *Ethnic and Racial Studies* 37, no. 10 (2014); Daniel Strand, "0.01%: Genetics, Race and the Methodology of Differentiation," *Eurozine*, January 4, 2021; Jenny Reardon and Kim TallBear, "'Your DNA Is Our History': Genomics, Anthropology, and the Construction of Whiteness as Property," *Current Anthropology* 53, suppl. 5 (2012); and Angela Saini, *Superior: The Return of Race Science* (Boston: Beacon Press, 2019).

52. Quoted from Wiktionary, where the Usage Note for "Mongoloid" in English reads: "Due to associations with old racial (and racist) theories (as with

Caucasoid, Negroid), and associations with Down syndrome, the term is highly offensive." Retrieved January 11, 2023.

53. We can recall here the debates around the HDGP covered in chapter 2. See also Reardon and TallBear, "'Your DNA Is Our History'"; Ruha Benjamin, *Race after Technology: Abolitionist Tools for the New Jim Code* (New York: Polity Press, 2019); and Hopman, "The Face As Folded Object."

54. Sommer, "Mirror, Mirror," 228.

55. Charlotte Hedenstierna-Jonson et al., "A Female Viking Warrior Confirmed by Genomics," *American Journal of Physical Anthropology* 164, no. 4 (2017).

56. Anna Källén et al., "Archaeogenetics in Popular Media: Contemporary Implications of Ancient DNA," *Current Swedish Archaeology* 27 (2019). See also Judith Jesch, "Haplotypes and Textual Types: Interdisciplinary Approaches to Viking Age Migration and Mobility," *Journal of Social Archaeology* 21, no. 2 (2021): 216–35.

57. Michael Greshko, "Famous Viking Warrior Was a Woman, DNA Reveals," *National Geographic*, September 12, 2017; Michael Price, "DNA Proves Fearsome Viking Warrior Was a Woman," *Science Magazine*, September 8, 2017.

58. Andreas Nyblom, "The Lagertha Complex," in *Critical Perspectives on Ancient DNA*, ed. Daniel Strand, Anna Källén, and Charlotte Mulcare (Cambridge, MA: MIT Press, 2024).

59. Erik Trinkaus and Pat Shipman, *The Neandertals: Changing the Image of Mankind* (New York: Knopf, 1993), 408. See also Amade M'charek, "Whitewashing the Neanderthal: Doing Time with Ancient DNA," in *Critical Perspectives on Ancient DNA*, ed. Daniel Strand, Anna Källén, and Charlotte Mulcare (Cambridge, MA: MIT Press, 2024).

60. Similar arguments connecting individual and collective identity are found in an article on the contemporary imagination and mediation of Viking identity by means of genetics: Marc Scully, Turi King, and Steven D Brown, "Remediating Viking Origins: Genetic Code as Archival Memory of the Remote Past," *Sociology* 47, no. 5 (2013): 921–38.

61. Quote from *The Viking Warrior Queen* (2020), cited in Nyblom, "The Lagertha Complex."

62. Oikkonen, *Population Genetics and Belonging*, 85.

63. See, for example, Chris Fowler, *The Archaeology of Personhood: An Anthropological Approach* (London: Routledge, 2004), for an account of the ethical and practical complexities of interpreting personhood from archaeological remains. See also Jerome De Groot, *Double Helix History: Genetics and the Past* (London: Routledge: 2022), chap. 3.

Chapter Five

1. *The Good Fight*, season 3, episode 2, featuring a conversation between characters Roland Blum and Maia Rindell, at around 30.00–31.00 min. (CBS All Access, 2019).

2. Donna Haraway, "Situated Knowledges: The Science Question in Feminism and the Privilege of Partial Perspective," *Feminist Studies* 14, no 3 (Autumn 1988): 575–99.

3. John Robb, "Towards a Critical Ötziography: Inventing Prehistoric Bodies," in *Social Bodies*, ed. Helen Lambert and Maryon McDonald (New York: Berghahn Books, 2009), 100–128, at 105.

4. Lynn Hasher, David Goldstein, and Thomas Toppino, "Frequency and the Conference of Referential Validity," *Journal of Verbal Learning and Verbal Behavior* 16, no. 1 (1977): 107–12.

5. Lisa K. Fazio et al., "Knowledge Does Not Protect against Illusory Truth," *Journal of Experimental Psychology* 144, no. 5 (2015): 993–1002.

6. Bryan Sykes, *The Seven Daughters of Eve* (New York: W. W. Norton, 2001), x.

7. Elizabeth D. Jones, *Ancient DNA: The Making of a Celebrity Science* (New Haven: Yale University Press, 2022), 189–90.

8. Howard Wolinsky, "Ancient DNA and Contemporary Politics," *EMBO Reports* 20 (November 7, 2019): e49507; Editorial, "Use and Abuse of Ancient DNA," *Nature* 555 (March 29, 2018): 559; Natan Elgabsi, "The 'Ethic of Knowledge' and Responsible Science: Responses to Genetically Motivated Racism," *Social Studies of Science* 52, no. 2 (2021): 303–23.

9. Anna Källén et al., "Archaeogenetics in Popular Media: Contemporary Implications of Ancient DNA," *Current Swedish Archaeology* 27 (2019). See also Joyce C. Havstad, "Sensational Science, Archaic Hominin Genetics, and Amplified Inductive Risk," *Canadian Journal of Philosophy* 52, no 3 (2022): 295–320.

10. For a similar argument, see Liv Nilsson Stutz, "Rewards, Prestige, and Power: Interdisciplinary Archaeology in the Era of the Neoliberal University," *Forum Kritische Archäologie* 11 (2022) (special issue, *Interdisciplinary Contentions in Archaeology*, ed. Artur Ribeiro and Alexandra Ion): 40–52.

11. In this vein, Harvard geneticist David Reich has described his own ambitions in terms of "build[ing] an American-style genomics factory" to enable the "industrial-scale study of ancient DNA." Quoted from David Reich, *Who We Are and How We Got Here: Ancient DNA and the New Science of the Human Past* (New York: Oxford University Press, 2018), xix.

12. There are innumerable variations on this theme to be found in popular reports of research on aDNA and human evolution. An excellent example is the award-winning National Geographic documentary, *Journey of Man: The Story of the Human Species*, featuring Spencer Wells and the Genographic Project.

13. Jennifer Raff presents a detailed and personal account of the laboratory procedure to extract ancient DNA in *Origins: A Genetic History of the Americas* (New York: Twelve Books, 2022) chap. 5.

14. E.g., Susanne E. Hakenbeck, "Genetics, Archaeology and the Far Right: An Unholy Trinity," *World Archaeology* 51, no. 4 (2019): 517–27; Catherine J. Frieman and Daniela Hofmann, "Present Pasts in the Archaeology of Genetics, Identity,

and Migration in Europe: A Critical Essay," *World Archaeology* 51, no. 4 (2019): 531–32.

15. E.g., Megan Gannon, "When Ancient DNA Gets Politicized," *Smithsonian Magazine*, July 12, 2019.

16. Johannes Krause, quoted from an interview with Howard Wolinsky, "Ancient DNA and Contemporary Politics," 2, 5.

17. David Reich, quoted from an interview with Howard Wolinsky, "Ancient DNA and Contemporary Politics," 2, 5.

18. E.g., "Use and Abuse of Ancient DNA: Researchers in Several Complementary Disciplines Need to Tread Carefully over the Shared Landscapes of the Past," editorial, *Nature* 555 (March 29, 2018): 559.

19. Jenny Reardon, *Race to the Finish: Identity and Governance in an Age of Genomics* (Princeton: Princeton University Press, 2004); Natan Elgabsi, "The 'Ethic of Knowledge' and Responsible Science: Responses to Genetically Motivated Racism," *Social Studies of Science* 52, no. 2 (2021): 303–23.

20. Marianne Sommer, " 'Wer sind Sie wirklich?'—Identität und Geschichte in der 'Gense-Quenz,'" *L'Homme* 21, no. 2 (2010): 51–70.

21. See also Anna Källén, "The Sigtuna Debacle: A Story of Ancient DNA, Immigrants and Fake News in a Viking Age Town in Sweden," in *Polarized Pasts: Heritage and Belonging in Times of Political Polarization*, ed. Elisabeth Niklasson (New York: Berghahn Books, 2023).

22. Susan Walsh, quoted in Colin Barras, "Ancient 'Dark-Skinned' Briton Cheddar Man Find May Not Be True," *New Scientist*, February 23, 2018.

23. There have been similar laudable attempts to engage in more transparent popular communication, for example, by the above-quoted geneticists Mark Thomas and Jennifer Raff, and by biological anthropologist John Hawks on his own blog, https://johnhawks.net/johnhawks/. There have also been important calls for more transparent communication in scholarly journals. See, for example, Marie-France Deguilloux et al., "European Neolithization and Ancient DNA: An Assessment," *Evolutionary Anthropology* 21 (2012): 24–37; John Hawks, "Accurate Depiction of Uncertainty in Ancient DNA Research: The Case of Neandertal Ancestry in Africa," *Journal of Social Archaeology* 21, no. 2 (2021): 179–96; and K. Ann Horsburgh, "Molecular Anthropology: The Judicial Use of Genetic Data in Archeology," *Journal of Archaeological Science* 56 (2015): 141–45. Horsburgh contributed also, together with geneticist-mathematician Hans-Jürgen Bandelt, in this vein to a Book Review Forum on *Who We Are and How We Got Here: Ancient DNA and the New Science of the Human Past* by David Reich in *Current Anthropology* 59, no. 5 (October 2018): 656–61. For a related discussion pertaining to forensic science, see Jason M. Chin and Carlos M. Ibaviosa, "Beyond CSI: Calibrating Public Beliefs about the Reliability of Forensic Science through Openness and Transparency," *Science & Justice* 62 (2022): 272–83.

24. Charlotte Mulcare and Mélanie Pruvost, "Found in Translation," in *Critical Perspectives on Ancient DNA*, ed. Daniel Strand, Anna Källén, and Charlotte Mulcare (Cambridge, MA: MIT Press, 2024).

25. Jerome De Groot, *Double Helix History: Genetics and the Past* (New York: Routledge: 2022), 8.

26. See also Ruha Benjamin, "The Emperor's New Genes: Science, Public Policy, and the Allure of Objectivity," *Annals of the American Academy of Political and Social Science* 661, no. 1 (2015): 130–42.

27. For a similar argument (albeit with a different conclusion regarding storytelling), see Patrick J. Geary, "Genetic History and Migrations in Western Eurasia, 500–1000," in *Empires and Exchanges in Eurasian Late Antiquity*, ed. Nicola Di Cosmo and Michael Maas (Cambridge: Cambridge University Press, 2018).

28. E.g., Laurent A. F. Frantz et al., "Animal Domestication in the Era of Ancient Genomics," *Nature Reviews Genetics* 21 (2020): 449–60; Oscar Estrada et al., "Ancient Plant DNA in the Genomic Era," *Nature Plants* 4 (2018): 394–96; Sebastián Duchêne et al., "The Recovery, Interpretation and Use of Ancient Pathogen Genomes," *Current Biology* 30, no. 19 (2020): R1215–31.

29. But see Venla Oikkonen, "Conceptualizing Histories of Multispecies Entanglements: Ancient Pathogen Genomics and the Case of *Borrelia recurrentis*," *Journal of Social Archaeology* 21, no. 2 (2021): 197–215, for an argument on the cultural and world-making implications of ancient pathogen genomics.

30. E.g., Daniela Hofmann, "What Have Genetics Ever Done for Us? The Implications of aDNA Data for Interpreting Identity in Early Neolithic Central Europe," *European Journal of Archaeology* 18, no. 3 (2015): 454; Krishna R. Veeramah, "The Importance of Fine-scale Studies for Integrating Paleogenomics and Archaeology," *Current Opinion in Genetics & Development* 53 (2018): 83–89; Martin Furholt, "Massive Migrations? The Impact of Recent aDNA Studies on Our View of Third Millennium Europe," *European Journal of Archaeology* 21, no. 2 (2018): 159–91; Susan E. Hakenbeck, "Genetics, Archaeology and the Far Right: An Unholy Trinity," *World Archaeology* 51, no. 4 (2019): 517–27; Thomas J. Booth et al., "Tales from the Supplementary Information: Ancestry Change in Chalcolithic–Early Bronze Age Britain Was Gradual with Varied Kinship Organization," *Cambridge Archaeological Journal* 31, no. 3 (2021): 379–400; Joanna Brück, "Ancient DNA, Kinship and Relational Identities in Bronze Age Britain," *Antiquity* 95, no. 379 (2021): 228–37.

31. E.g., Maïté Rivollat et al., "Ancient Mitochondrial DNA from the Middle Neolithic Necropolis of Obernai Extends the Genetic Influence of the LBK to West of the Rhine," *American Journal of Physical Anthropology* 161, no. 3 (2016): 522–29; Carlos Eduardo G. Amorim et al., "Understanding 6th-century Barbarian Social Organization and Migration through Paleogenomics," *Nature Communications* 9, no. 3547 (2018); Alissa Mittnik et al., "Kinship-based Social Inequality in Bronze Age Europe," *Science* 366 (2019): 731–34; Chris Fowler

et al., "A High-resolution Picture of Kinship Practices in an Early Neolithic Tomb," *Nature* 601 (27 January 2022).

32. Interdisciplinary collaboration has been a recurring theme in discussions of aDNA research, and there have been a number of attempts to formulate best practice documents and guidelines for further successful collaboration between geneticists, archaeologists, and other stakeholders. For a critical discussion and examples, see Mulcare and Pruvost, "Found in Translation"; and Anna Källén et al., "Petrous Fever: The Gap between Ideal and Actual Practice in Ancient DNA Research," *Current Anthropology* (2024).

33. Donna Haraway, *Staying with the Trouble: Making Kin in the Chthulucene* (Durham, NC: Duke University Press, 2016), 12.

34. Rachel J. Crellin and Oliver J. T. Harris, "Beyond Binaries: Interrogating Ancient DNA," *Archaeological Dialogues* 27, no. 1 (2020): 37–56.

35. E.g., Erika Balsom, "There Is No Such Thing As Documentary: An Interview with Trinh T. Minh-ha," *Frieze Los Angeles* 199 (November 1, 2018); Trinh T. Minh-ha, in *Heritage and Borders*, ed. Anna Källén (Stockholm: Royal Swedish Academy of Letters, History and Antiquities, 2019), 151, 158–59.

36. José van Dijck, *Imagenation: Popular Images of Genetics* (London: Macmillan, 1998); Siddhartha Mukherjee, *The Gene: An Intimate History* (London: Bodley Head, 2016). Compare Harden, *The Genetic Lottery*.

37. See also Crellin and Harris, "Beyond Binaries."

Index

Page numbers in italics refer to figures.

admixture, 39, 41, 67, 70, 73–74, 82, 86, 139n54
African Americans, 65
American Journal of Physical Anthropology, 1
Amnesty International, 83
Amstutz, Ruth, 76
ancestry, 19; as capital, 62–64; and genetic tests, 62–65, 75–76, 81; and kinship, 81, 85. *See also* ancient DNA (aDNA); genetics
"Ancient Beringian" population, 40–41, 45, 49–50
ancient DNA (aDNA), 7–11, 17–23, 44–45, 51–53, 59–61; and ancestry, 65–66; critical debates around research, 73, 115; fragmentary, 105; human, 19, 31, 48; illusory truth effect of, 110–11; interdisciplinary collaboration on, 151n32; modeling populations from, 49; percentage of DNA extracted and sequenced from, 129n42; as petrified book of life, 30–31; and race, 82; "revolution" in, 26–32, 94; transparency in research, 115–16; and value of ancestry, 77–82; "vanishing resource" challenge

of, 73–74. *See also* ancestry; ancient human remains; DNA; genomic science; high-throughput sequencing
ancient human remains, 17–18, 61, 73–74. *See also* ancient DNA (aDNA)
Ancient One (Kennewick Man), 100
Anderson, Benedict, *Imagined Communities*, 47
Anzick-1, *41*, 49
archaeogenetics, 19, 26, 30, 81, 101, 117
archaeology, 6–7, 28, 35–38, 60; culture-historical, 36, 38, 131n7; storytelling in, 4
arrow, 35–60, 110, 131n4
"Aryans," 38, 50–51, 60, 114

Bering, Vitus Jonassen, 49
Bhabha, Homi, 4
big data, 24, 28, 30
bioethics, 67, 73
bioethnic diversity, 70, 76–77
bioinformatics, 24–28, 30
black box, 30, 44, 115, 126n13
Blair, Tony, 23–24
blood, 17, 65–67
Brexit, 93–95
Bridget, Saint, 61

153